HUMAN ANATOMY
AND PHYSIOLOGY

HUMAN ANATOMY AND PHYSIOLOGY

David Le Vay, MS, FRCS

TEACH YOURSELF BOOKS

Long-renowned as the authoritative source for self-guided learning – with more than 30 million copies sold worldwide – the *Teach Yourself* series includes over 200 titles in the fields of languages, crafts, hobbies, sports, and other leisure activities.

British Library Cataloguing in Publication Data

Le Vay, David
   Human anatomy & physiology. — 3rd ed.
   —(Teach yourself books)
   1. Human physiology
   I. Title   II. Series
   612 QP34.5

ISBN 0 340 42875 9

Library of Congress Catalog Card Number: 93–83167

First published in UK 1974 by Hodder Headline Plc, 338 Euston Road, London NW1 3BH

First published in US 1993 by NTC Publishing Group, 4255 West Touhy Avenue, Lincolnwood (Chicago), Illinois 60646 – 1975 U.S.A.

Copyright © 1974, 1985, 1988 Hodder & Stoughton

Printed in Great Britain by Cox & Wyman Ltd, Reading, Berkshire.

First published 1974
Second edition 1985
Third edition 1988

Impression number	15	14	13	12	11	10	9
Year	1999	1998	1997	1996	1995		

Contents

structures of the leg. Muscles of the upper leg. Muscles of the lower leg. The blood vessels of the leg. The nerves of the leg.

Acknowledgements

I should like to thank the following for permission to reproduce illustrations and material: Longmans, Green & Co., as regards *Gray's Anatomy* and J. & A. Churchill in respect of their *Principles of Human Physiology*. I have also drawn widely from that invaluable work of reference, *A Companion to Medical Studies, Volume 1*, ed. R. Passmore and J. S. Robson, published by Blackwell Scientific Publications, and the figures reproduced therefrom are individually acknowledged in the text.

Introduction

Anatomy

The word *anatomy* means the cutting up of the body to examine its parts. Knowledge gained in this way is essentially *regional* or *topographic*; one gains a familiarity with each part, such as the arm or leg. However, every part contains the same *kinds* of organs – blood vessels, nerves, bones, and so on – so that, superimposed on regional anatomy, there is also a *systemic* aspect, in which the body is considered to be made up of several coordinated systems: vascular, nervous, skeletal and so on. Gross observations of this kind are known as *macroscopic* anatomy, and contrasted with *microscopic* anatomy, or *histology*, the study of the fine structure of cells and tissues.

All this implies a static view of adult anatomy. However, we have also to consider developmental anatomy: consisting first, of *embryology*, the growth of the individual within the womb from the single cell formed by the fusion of ovum and spermatozoön, and, secondly, the *postnatal* development from infancy to maturity. And towards the end of life certain changes of *senescence* appear.

The development of the individual is known as *ontogeny*, as contrasted with *phylogeny*, the development of the race from more primitive forms. To some extent the individual recapitulates phylogeny during his own development. *Comparative anatomy* is the study of the human body in relation to that of animals; we sometimes use the word *morphology* for the study of differences and resemblances of corresponding organs. For instance, the human foot and the horse's hoof are basically similar structures that have evolved on differing lines.

Methods

The oldest method of acquiring anatomical knowledge is by *dissection*, a process extended later by *microscopy*. But we need to check this knowledge against that obtainable from the living body.

Surface anatomy deals with the relation of superficial landmarks to the deeper structures. Inspection reveals the bulge of muscles when they contract, the pulsation of arteries and the course of veins, and the position of bony prominences.

Palpation, manipulation and *percussion* reveals respectively the consistence of the deep structures, the movements of joints, and the boundaries of hollow, air-containing or solid organs, while listening with a stethoscope – *auscultation* – locates organs such as the heart, lungs and bowel. *Endoscopy* is the introduction of an instrument for visualising the inside of structures such as the ear or stomach with the eye or the camera, and *surgical exploration* is a useful source of information.

In *radiographic anatomy*, X-rays show the skeletal system and also demonstrate hollow organs when these are filled with substances opaque to the rays. *Cineradiography* demonstrates the movement in life of the joints, heart, lungs and viscera. Radiographic anatomy is valuable because the positions of organs such as the stomach vary greatly with changes of posture and emotional disturbance.

Variation

Human beings are essentially similar, but there is a continuous variation from the standard pattern in inessential details. We vary outwardly in height and colouring, and internally there may be minor aberrations in the arrangement of nerves and blood vessels, bile-ducts or bronchi. But just as the general catalogue – 'item, two lips, indifferent red; item, two eyes, with lids to them; item, one neck, one chin and so forth – is always correct, so the femoral artery or the biceps muscle is always where we expect to find it. There are occasional grosser errors which represent a failure of normal developmental processes such as the absence of part of a limb, or the transposition of heart or viscera.

Introduction

Anatomy

The word *anatomy* means the cutting up of the body to examine its parts. Knowledge gained in this way is essentially *regional* or *topographic*; one gains a familiarity with each part, such as the arm or leg. However, every part contains the same *kinds* of organs – blood vessels, nerves, bones, and so on – so that, superimposed on regional anatomy, there is also a *systemic* aspect, in which the body is considered to be made up of several coordinated systems: vascular, nervous, skeletal and so on. Gross observations of this kind are known as *macroscopic* anatomy, and contrasted with *microscopic* anatomy, or *histology*, the study of the fine structure of cells and tissues.

All this implies a static view of adult anatomy. However, we have also to consider developmental anatomy: consisting first, of *embryology*, the growth of the individual within the womb from the single cell formed by the fusion of ovum and spermatozoön, and, secondly, the *postnatal* development from infancy to maturity. And towards the end of life certain changes of *senescence* appear.

The development of the individual is known as *ontogeny*, as contrasted with *phylogeny*, the development of the race from more primitive forms. To some extent the individual recapitulates phylogeny during his own development. *Comparative anatomy* is the study of the human body in relation to that of animals; we sometimes use the word *morphology* for the study of differences and resemblances of corresponding organs. For instance, the human

1

Cells and Tissues

Cells

The unit of living tissue is the microscopic *cell*; even the fertilised ovum, the largest cell, is only just visible to the naked eye. The body is composed of different *tissues*, each of which is an aggregation of similar cells, together with intercellular substance. In addition, material is produced by cell activity, which is no longer living, e.g. the mineral content of bone, the hair and nails. All the body cells have sprung from the division of the female germ cell (ovum) after fertilisation by the male germ cell (spermatozoön).

In the adult there are specialised cells that cannot divide and are irreplaceable, e.g. nerve and muscle cells; cells that divide slowly in health but can be stimulated to rapid growth in case of need, e.g. connective tissue in repair after injury; and cells that reproduce rapidly to replace those with a short life-span, e.g. the precursors of red blood cells and the epithelium of the skin. Many cells have bizarre shapes. Some nerve cells, though minute, send their fibres almost the length of the body. Muscles are made up of elements often as long as the muscle itself. The red blood cell is modified into a flattened biconcave disc to concentrate the blood pigment (haemoglobin) at the periphery.

The essential parts of any cell are the *bounding membrane*, the protoplasmic substance, or *cytoplasm*, and the *nucleus*. The bounding membrane is permeable to simple ions. The cell *surface* has characteristics of the tissue, the species and the individual; it also contains proteins which reject foreign material from other persons

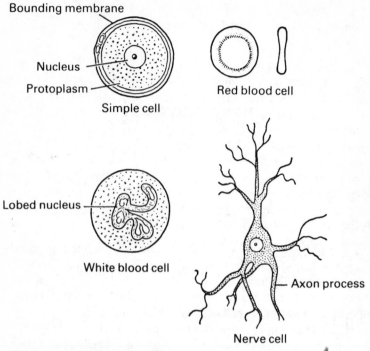

Fig. 1.1   Various types of cells

or animals – i.e. there is an *immunological* function. The *cytoplasm* is all the material around the nucleus. It is a mass of colloidal proteins, carbohydrates and solutions of smaller molecules, and contains most of the ribonucleic acid (RNA) of the cell. It also contains a number of discrete arrangements, or *organelles*, of various kinds:

1 The *Golgi apparatus*, near the nucleus, function unknown.
2 The *mitochondria*, thread-like bodies – thousands in a cell which contain enzymes for the oxidation of carbohydrates. They are the main sites for energy conversion, and are prominent in very active tissues such as heart muscle and kidney tubules.
3 The *lysosomes* contain enzymes concerned with digesting absorbed material and scavenging.
4 The *ribosomes*, dense particles of RNA-protein complexes; points of production of protein molecules.

5 The *centrioles*, a pair of barrel-shaped bodies at right angles to each other near the nucleus, which have an important function in cell division.

The *nucleus* is sharply defined and can be separated out of the cell. It contains the *chromosomes* (see p. 8) with RNA and long molecules of DNA (deoxyribonucleic acid). There are also small *nucleoli*, aggregates of RNA. Certain very large cells are multi-nucleate, e.g. bone osteoclasts and skeletal muscle fibres. The nuclei of the polymorph white blood cells are lobulated.

Cells stick together and this facilitates their arrangement in tissues. Cancer cells lose this adhesion and then invade the body. Some cells have the power of *phagocytosis*, i.e. they can surround and ingest foreign matter, bacteria and dead cells; this is notable in the polymorphs of the blood and the cells of connective tissue. The chemical constituents of cells include proteins – some in colloidal solution in the cytoplasm, some particulate in the organelles – droplets of fats or lipids, and carbohydrates as soluble simple

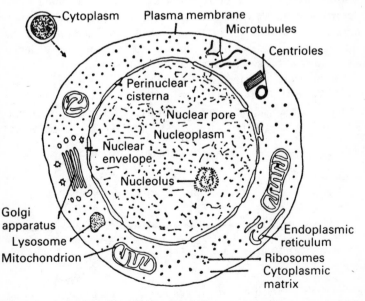

Fig. 1.2 A section of a cell containing the structures that are essential for its survival (From *A Companion to Medical Studies, Vol. 1*)

compounds or granules of complex polysaccharides. There are also *pigments*: the red haemoglobin of blood; its pink derivative in muscle; its green and brown degradation products in bile and faeces; and the photosensitive visual purple of the retina.

The cells of different tissues may be greatly modified for their functions, e.g. the red blood cell has lost its nucleus, the nerve cell has an enormously elongated axon process to carry stimuli, and the striated muscle cell can contract in length. Despite these modifications, the essential components remain recognisable in each case.

Cell division

Every cell in the body (except the germ cells) carries within its nucleus the forty-six chromosomes in twenty-three pairs characteristic of man; there are twenty-two pairs not involved in sex determination – the *autosomes* – and one pair of *sex chromosomes*. These are not clearly visible until the preliminary phase of cell division. Each DNA molecule contained in the chromosomes consists of a long pair of spirally-wound strands – the double helix. This consists of a backbone of deoxyribose (sugar) and phosphate molecules with purine or pyrimidine bases attached – adenine, thymine, guanine and cytosine. The bases on one strand face those of the other in precise and regularly repeated fashion, and the order of these pairs of bases contains the information or genetic code controlling protein synthesis, ensuring that the constituent amino-acids are inserted in the correct places in the growing polypeptide chains. The organisation is complex, and is effected via 'messenger' RNA and 'transfer' RNA molecules. The DNA of the cell is not

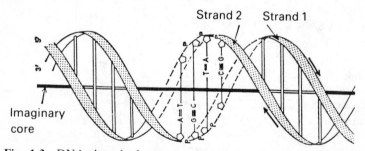

Fig. 1.3   DNA thread of two strands round an imaginary core in an anti-parallel manner; A, T, G and C represent complementary base parts (From *A Companion to Medical Studies, Vol. 1*)

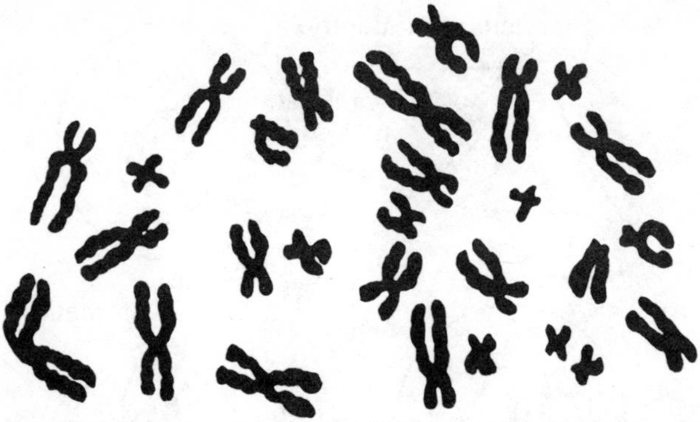

Fig. 1.4 Chromosomes of a normal human male cell from a culture of peripheral blood (Adapted from *A Companion to Medical Studies, Vol. 1*)

confined to the nucleus. Some is in the cytoplasm, concerned with protein synthesis, but that of the nucleus is contained in the separate chromosomes.

Cell division in man is an elaborate process of *mitosis*, designed to ensure exact duplication of the original cell and exact sharing of the genetic material, for it is this material that contains the information required for the synthesis of new cells (Fig. 1.5).

In the resting *interphase*, the chromosomes are not visible as separate structures. As mitosis approaches, they become visible as threads within the nucleus (*prophase*). The centrioles divide and move to opposite poles of the cell, from which radiates a controlling spindle of protein fibres. The nuclear membrane disappears. The chromosomes are now seen to be double and align themselves in the plane of the spindle between the centrioles. The two halves of each chromosome separate and are pulled by contracting spindle fibres to opposite ends of the cell to form the new chromosomes of the two daughter nuclei. The cytoplasm of the cell cleaves in two and new nuclear membranes form around each separated bundle of chromosomes, reconstituting two daughter cells and two daughter nuclei genetically identical to the parent nucleus.

Tissues

There are only four essential tissues: epithelium, connective, muscular and nervous. These may be modified in various ways, but

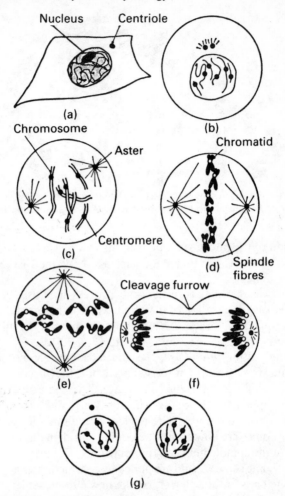

Fig. 1.5 Mitosis: (a) Interphase (forty-six chromosomes); (b) Early prophase. Centriole divides, chromosomes appear as threads; (c) Late prophase. Centrioles separate. Aster and spindles form, chromosomes seen to be double, nuclear membrane disappears. (d) Metaphase. Chromosomes line up in equatorial plane. (e) Early anaphase. Centromeres divide, chromatids begin to separate. (f) Late anaphase. Separation of chromatids continues, cleavage furrow forms. (g) Telophase. Two daughter cells each with forty-six chromosomes (From *A Companion to Medical Studies, Vol. 1*)

together they compose the structure of the body. They differ in the nature of their component cells and the intercellular ground substance or *matrix*.

Connective tissue

This is the framework of the body. It is characterised by the large amount of intercellular substance, and the exact nature of a connective tissue depends on its matrix and the cells and fibres it contains. It includes widely different tissues: bone, cartilage, fibrous tissue, elastic tissue, blood and lymph.

In simple *areolar* tissue – the loose packing in clefts and spaces, as in the *subcutaneous layer* between skin and deeper structures – there is a semi-fluid matrix containing mucopolysaccharides in which run strands of protein fibres – white (unyielding) and yellow (elastic) – together with rather sparse cells, an arrangement allowing free gliding. This tissue contains much of the *extracellular fluid* of the body. An excess of white fibres is found in the firm fibrous tissue of tendons and ligaments and sheets of membrane (*fasciae*).

Adipose connective tissue

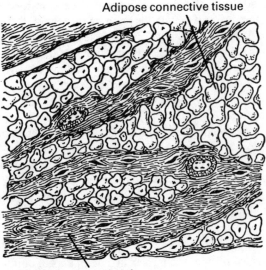

Areolar connective tissue

Fig. 1.6   A simple connective tissue

Elastic fibres predominate where stretch and recoil are important, as in the arteries. *Adipose* tissue is a store of triglycerides and a source of energy. It insulates against cold and buffers the internal organs, and is more developed in women. It may disappear almost entirely in starvation. It is divided into lobules by fibrous septa and is closely packed with cells, each of which has its nucleus at one side and is distended by a fat globule.

In *blood* and *lymph* the matrix is fluid, while the cells are modified to carry dissolved gases or deal with invading micro-organisms. In *cartilage* and *bone* the matrix has become hard by impregnation with mineral salts.

In general, the cells of connective tissue fall into three types: *fibroblasts*, which manufacture fibres; *mast cells*, concerned with carbohydrate metabolism; and *phagocytes*, which scavenge foreign material and the debris of dead cells, and may be mobile.

Epithelium

Epithelia are sheets of cells covering body surfaces: the external aspect of the skin and the internal surfaces which line the body cavities. In addition, they line the vessels, the respiratory, digestive and urinary tracts, and the glands opening into them. Both internal and external surfaces of the body possess this intact epithelial envelope, as a barrier between the outside world (including, para-doxically, the contents of the bowel and hollow organs) and the deeper intermediate structures. These inner and outer linings blend at the orifices of the various cavities – at the lips, anus, nose and urethral apertures, where the damp linings or *mucous membranes* of the digestive, respiratory and urinary tracts become continuous with the skin.

Epithelium is very cellular and simple in arrangement, with little ground substance. The simplest form is one cell thick, with a basement membrane. It may be *squamous*, arranged in mosaic fashion, as at the skin surface, or *stratified* in row upon row of cells.

Glands

These consist of simple or complex arrangements of epithelial cells forming useful substances in secretory cells. The simplest form is the *goblet cell* of the intestinal mucous membrane. But they may be more complex: tubular – simple, branched or coiled – or com-

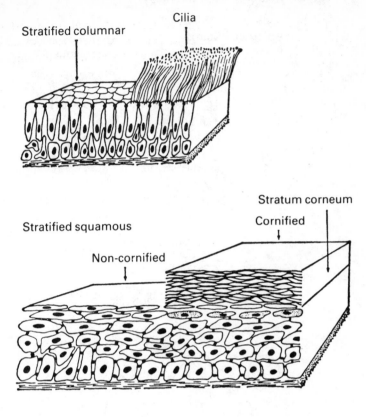

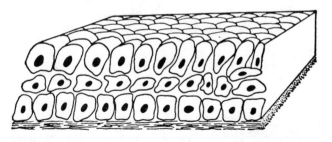

Fig. 1.7 The stratified epithelia (From *A Companion to Medical Studies, Vol. 1*)

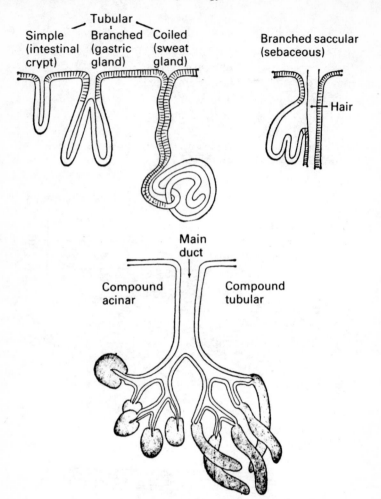

Fig. 1.8   Types of glands (Adapted from *A Companion to Medical Studies, Vol. 1*)

pound, with a branching duct system ending in blind pouches, or *acini*. Acini are either mucous or serous. The cells of mucous acini appear empty; their clear cytoplasm contains polysaccharides. The serous acinar cells are granular, have a large nucleus and manufacture protein enzymes. Glands may also produce salt and solute (sweat glands), acid (gastric mucosa), alkali (duodenal mucosa) or grease (subcutaneous glands). They may secrete continuously or only in response to some stimulus, such as the entry of food into the stomach. They may be independent or under nervous, chemical or hormonal control. If they discharge their secretion at the skin surface or into an internal organ, they are called *exocrine*. If they discharge directly into the bloodstream, they are *endocrine* glands and their *hormones* exert a remote influence on target organs at a distance.

It is epithelium that gives rise to the hair, nails and teeth; *endothelium* is the name given to the smooth linings of blood vessels; the glistening layers that line the great body cavities, pleural and peritoneal, are sometimes called *mesothelia*; and a *carcinoma* is a form of cancer in which epithelial cells have lost their adhesion and invaded the subjacent tissues.

2

Basic Biophysics and Biochemistry

Homeostasis

The human body is exposed to a constantly changing *external* environment. These changes are neutralised by the *internal* environment – the blood, lymph and tissue fluids that bathe and protect the cells. This stability is the object of the vital mechanisms. Let us take some examples.

Exercise produces heat and the body temperature tends to rise; then perspiration occurs, and there is heat loss due to the evaporation of water which compensates for the raised temperature – i.e. sweating is a physiological cooling mechanism that operates in exertion or fever.

Again, the blood is normally slightly alkaline. Although during exercise the muscles produce carbon dioxide, there is little or no rise in its acidity, firstly because of its chemical buffering properties, and secondly because the excess CO_2 is exhaled as it is formed; hence rapid breathing during effort and subsequent panting as the oxygen debt is paid off.

A third instance of homeostasis is that the blood level of sugar (glucose) remains constant during fasting, even though sugar is being burnt by the tissues, because sugar is formed from fat and protein in the body stores, mainly in the liver.

Finally, most of the water drunk enters the bloodstream; yet even large quantities of fluid do not lower the osmotic pressure of the blood by dilution. Any tendency in this direction is sensed by special *osmoreceptors* at the base of the brain which affect the output of a pituitary hormone that controls the output of water by the kidney. Excretion of water in the urine keeps pace with absorption.

These examples illustrate how the internal environment is kept constant despite external stresses. Were these mechanisms to fail, the organism would be in danger, e.g. in high fever due to infections, or when the blood sugar falls to a low level.

Body water

Water is the universal fluid medium of the body. It transports food, wastes and respiratory gases, and allows them to diffuse in and out of the cells. Water constitutes up to 95% of cell protoplasm and 60% of body weight, and no substance can influence living cells until it is brought into watery solution.

The *total body water* is some forty litres in a normal healthy adult man, or two-thirds of the body weight. This includes the water in free fluids, like the blood plasma and the fluid of the tissue spaces, as well as that within the cells. The *extracellular water* is about fifteen litres, or one-fifth of the body weight. Three litres are contained in the plasma; twelve litres in the interstitial fluid. The remaining *intracellular water* amounts to twenty-five litres:

Total body water 40 l
$\begin{cases} \text{Intracellular 25 l} \\ \text{Extracellular 15 l} \end{cases}$
$\begin{cases} \text{Interstitial fluid 12 l} \\ \text{Plasma 3 l} \end{cases}$

The total content of extracellular fluid (ECF) varies a little and so the body weight in the morning (after urination) fluctuates by up to a kilogramme. The ECF is reduced by lack of water intake, or by increased output due to sweating, vomiting or diarrhoea. It is increased in waterlogging of the tissues (oedema), as in starvation, heart failure or kidney disease.

There are important differences in the composition of the ECF and the intracellular fluid (ICF). The *cations* of the ECF are mainly sodium, a little potassium, calcium and magnesium; those of the ICF are mainly potassium, some sodium and calcium, and rather more magnesium. The *anions* of the ECF are mainly chloride, with some bicarbonate; those of the ICF include some chloride, bicarbonate and sulphate, but are mainly organic acids, phosphate and proteins. Thus the principal difference is that sodium is the chief cation outside the cells and potassium inside; and this difference is essential and constant. The ECF, or internal environment, is mainly a solution of NaCl. The bicarbonate of the ECF is controlled by respiration, which expels CO_2 and regulates the acidity of the fluid.

Body water is derived from the water content of solid food, the water drunk and the *metabolic water* formed in the tissues by oxidation of the hydrogen in the food. Water is lost in the urine, in faeces, and by evaporation from the skin and lungs. The first two may vary greatly, but the expired air is always saturated with water vapour. The *water balance* in an adult sedentary male in a temperate climate is as follows:

Intake: food 800 g; drink 1300 g; metabolic 250 g = 2350 g.
Output: urine 1500 g; faeces 25 g; evaporation 825 g = 2350 g.

The evaporated water is important in cooling the body and accounts for a quarter of total heat loss. Because water loss from skin and lungs is obligatory, and because there is little or no reserve of water, a negative balance is easily produced if intake falls, if there is excess loss in diarrhoea and vomiting, if there is excess sweating, if there is excess loss from expiration in cold, dry conditions or if kidney disease eliminates excess water. Death occurs in less than a week if an individual is deprived of water. On the other hand, it is not possible to create a positive balance by drinking large amounts, for the excess water is lost in urine.

From the *biophysical* angle, bodily function is concerned with the behaviour of molecules in aqueous solution. Thus, by diffusion, the salts in the blood or the sugar content of the cerebrospinal fluid are uniformly concentrated.

Certain membranes allow water and dissolved substances to pass freely; if such a membrane separates two solutions of different strengths, their concentration is equalised by the transfer of water and solute in reverse directions. This is called *dialysis*. Membranes vary in the size of the molecules they allow to pass, and most cell membranes are only semi-permeable; allowing a free passage to water but only permeable to some solutes. If a semi-permeable membrane separates two solutions of a substance that cannot traverse the membrane, water is drawn from the weaker to the stronger solution. This attraction of water is known as *osmosis*. The *osmotic pressure* of the solution is the expression of the energy gradient set up by the tendency of water always to dialyse, so as to dilute the stronger solution.

The behaviour of living cells is intimately affected by the relative osmotic pressures of the surrounding medium and of the cell substance, for the intervening cell membrane is semi-permeable. A red blood cell, immersed in strong salt solution, shrivels as water is sucked out of it, but swells up and bursts in a weaker solution or in water.

There is obviously an osmotic pressure for the surrounding medium which is identical to that of the protoplasm of the cell; such a solution leaves the cell unaffected and is known as *isotonic*. The stability of the cells is such that their osmotic pressures are identical and isotonic with a 0·9% solution of sodium chloride, or *normal saline*. Normal saline is thus identical osmotically to the blood and lymph, and is used to bathe cells and tissues without harming them. It can also be given as an intravenous infusion to restore a depleted blood volume, as in surgical shock.

Whatever the substance, solutions containing the same number of molecules per unit volume have the same osmotic pressure. Therefore, solutions of different substances of the same *percentage* strength have osmotic pressures inversely proportional to the sizes of their molecules – e.g. a 1% solution of protein, which has complex molecules, contains far fewer particles per cubic centimetre than a 1% solution of sugar, and its osmotic pressure is correspondingly less. Further, many substances called electrolytes *ionise* in solution, i.e. their molecules split or dissociate into two or more electrically charged *ions*. The osmotic pressure of such solutions is greater than would be expected from the size of the molecules, for it is proportional to the total number of *particles*.

Solutions have a further property of physiological importance due to this tendency to ionise. This property is their *chemical reaction*: acid, alkaline or neutral. Acidity is due to the presence of hydrogen ions (H), and alkalinity to hydroxyl ions (OH); the reaction of water is neutral because it produces these ions in equal amounts. In any solution the product of hydrogen and hydroxyl ions is constant, the two being inversely proportional. So the acidity or alkalinity of a solution may be estimated by measuring its hydrogen-ion concentration alone. This is always a small figure (a litre of water contains only 10 g of such ions), and it is customary to use the power of 10 as a positive number – the hydrogen-ion exponent or pH.

Thus the pH of water is 7, and this is the centre point of chemical neutrality. The pH of a solution falls as its acidity rises, and increases as it becomes alkaline, a single step from, say, pH 7 to 8, indicating a tenfold increase of alkalinity. The reaction of the blood is normally just on the alkaline side, at pH 7·4. Cell functions are profoundly modified by minor changes in chemical reaction; a small increase in acidity or alkalinity may bring the heart to a stop. Nevertheless, small amounts of acid and alkali are constantly being formed in life, and the effects are minimised by certain *buffer salts*, such as sodium bicarbonate, in the body fluids and protoplasm which take them up automatically to prevent any change in reaction. The main buffer in the body fluids is sodium phosphate, and in the cells, sodium bicarbonate; each can vary between its own acid and alkaline forms to compensate for any change in the reaction of the medium. Proteins can function either as weak acids or as bases, and so stabilise the pH of the cells they compose.

Crystalloids and colloids

All substances fall into two classes: *crystalloids*, simple crystallisable compounds like sugar and salt which diffuse easily in water and traverse animal membranes; and *colloids*, complex materials like gelatin and egg-white which crystallise with difficulty or not at all, diffuse slowly and cannot cross membranes. The two can be separated by the process of dialysis, and the essential difference between them lies in the size of their molecules; the molecular weight of salt is 58·5; that of proteins is of the order of 100000.

Colloids may exist either as *sols* in apparent solution or as *gels*, a solid jelly state with water. Colloidal 'solutions' are really only suspensions of particles, a disperse phase in some fluid medium. These so-called phases are reversible, as when milk, a suspension of oil in water, becomes butter, a suspension of water in oil.

One feature of colloidal 'solutions' is that their surface layer has a much greater concentration of colloid than the mass of the solution; this is characteristic of the bounding membrane of living cells, the membrane whose semi-permeability is responsible for osmotic transfers, and for maintaining certain differences between the contained protoplasm and the surrounding medium.

Thus there is more potassium than sodium in red blood cells, though the reverse is the case in the plasma. Again, most cells have a slightly acid reaction of pH 6·8, although immersed in neutral or slightly alkaline media. These differences are maintained by the selective action of the cell membrane. But this selective action is not purely physical, for the outer layer contains fatty materials and can be penetrated by agents soluble in fats, such as ether and alcohol (anaesthetics act by influencing the brain cells in this way). Colloidal solutions are unstable, and the particles may be precipitated by heat, by changes in reaction or the addition of salts; the *coagulation* of protein sols by heat is familiar in the albumin of egg-white. A further important property of colloids is their power of *adsorption* of other substances due to the immense surface area presented by the dispersed particles. Adsorption is the capacity to pick up and retain a substance without entering into chemical combination with it.

The basic physical factors that control the transference of substances across cell membranes are:

1 A difference of *hydrostatic pressure* between the two sides. Thus the osmotic pressure of the proteins dissolved in the blood plasma resists the outward passage of water from the capillaries into the kidney tubules. This pressure is the equivalent of 30 mm of mercury. It is normally overcome by the hydrostatic effect of the blood pressure – say, 130 mm of mercury – which drives the water across. But if the blood pressure falls below 30 mm, because of disease or shock, urine ceases to be formed.

2 The ordinary laws of *diffusion* and *osmosis*, depending on the relative strengths of the solutions on either side.

3 Differences of *electric potential*.

4 The varying *permeabilities* of the membranes. Only rarely are they absolutely impermeable, and none is completely semi-permeable, allowing passage of water only. Some allow certain colloids to cross, more often water and crystalloids only. In certain cases, substances only pass in one direction, due to changes they produce in the membrane. Finally, in the dialysis of a compound of one diffusible and one non-diffusible ion, a difference of chemical reaction develops on the two sides of the membrane; this is the basis for the secretion of the highly acid and alkaline digestive juices of the stomach and pancreas respectively.

Energy transformations within the body

The essential chemical elements of protoplasm are carbon, hydrogen, nitrogen, sulphur and phosphorus. Simple compounds of these, such as water and carbon dioxide, require an elaborate conversion into organic matter, which is effected by green plants under the influence of sunlight. Neither man nor animals can achieve this; they depend on the consumption of vegetable matter, either directly or indirectly, after it has been utilised by other animals which serve as food in their turn. Ultimately, all flesh is grass. The energy transformations of living matter begin with light absorbed by plants and end with the heat waste of plants and animals. This energy passage is subject to the laws of thermodynamics and can be utilised to do *work*. (Note that the living organism has temporarily reversed the universal tendency for energy to run down and dissipate as heat.)

During digestion and absorption, the constituents of the food are broken down into simpler organic substances, which are reassembled in the tissues. The finished products differ slightly in composition in man and animals, and to some extent between one man and another. Man depends for body-building material and sources of energy on the complex organic compounds – fats, carbohydrates and proteins – that have been built up in living plants. The only chemical element he takes up in the free state is the oxygen in the inspired air. Oxygen frees the potential energy of the food by burning it in the cells of the body to form carbon dioxide and water, and liberating energy. Thus this last step is the complete reversal of what the green plant originally achieved. Since the breakdown products of animal tissues are used again by plants, there is a continuous cycle, on which both plants and animals depend. And the energy for our body heat and movement is ultimately derived from the sun.

Energy (sunlight) + water + carbon dioxide = organic matter

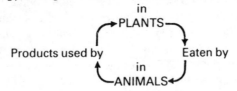

Organic matter + oxygen = carbon dioxide + water + heat and energy

The energy concentration of the green leaf is very low, hence the large amounts eaten by herbivores. Nevertheless, this energy can be concentrated by herbivores into fat and meat for use in human diet.

This cycle is only concerned with carbon, hydrogen and oxygen, but similar cycles apply to the other essential elements, such as nitrogen. This is mainly synthesised into organic matter by the soil bacteria on which plants depend. The plants pass it on to man and animals, which return it in the urea and ammonia of their excreta to the soil. A number of metallic ions essential to the body are taken in the form of simple mineral salts: sodium, potassium, calcium, magnesium and iron. Essential basic radicles are phosphates and chlorides.

Most natural foods contain little *sodium*; hence the importance of salt in the diet. Although a healthy individual does not retain excess salt, the content of the body is maintained well above that of the environment by the kidney. Sweat contains salt, and severe salt loss may occur in hot or humid conditions and cause cramps.

Potassium is abundant in plant and animal cells, and dietary deficiency is rare. Severe potassium loss may occur in vomiting and diarrhoea, and causes weakness and thirst.

Calcium is essential for the formation of bone during growth and the maintenance of bone mass in the adult, since there is a constant turnover of the kilogramme of calcium contained in the skeleton. Most of the calcium in the diet is lost in the faeces as insoluble salts; the problem is one of absorption. Milk and cheese are both rich sources. The calcium of bread is more readily absorbed in white bread, for in wholemeal the phytic acid in the husk forms insoluble calcium phytate which is excreted and so it is reasonable to enrich high-extraction bread with calcium lactate. Calcium is necessary for normal electrochemical excitability of tissues; deficiency can cause twitching and spasm of the muscles, known as *tetany*.

Magnesium is the most important intracellular cation after potassium, and is essential in certain enzyme reactions.

Iron is an essential mineral, as part of the molecule of the red pigment of the blood, haemoglobin, responsible for transferring oxygen from the lungs to the tissues. The body contains 4–5 g of iron, half of it in the haemoglobin. It is in constant demand for synthesis of the pigment in new red cells. A good deal is lost at each menstruation, and it is needed in pregnancy for the foetus and later for the milk. Hence the iron requirements of a woman of child-bearing age are twice those of a man, and iron-deficiency anaemia is more common in women.

Most of the dietary iron is lost in the faeces, and the amount absorbed is only just adequate for women. Rich sources are liver, offal, meat, fruit and vegetables, and iron may be absorbed from cooking pots. Oddly, the body is quite unable to excrete iron, and an excessive intake may lead to iron deposits in the tissues.

Lastly, there is a group of *trace elements*, which, although present in the tissues in minute quantities, are essential dietary constituents. These include iodine, fluorine, copper, cobalt, zinc and manganese.

Iodine is an essential constituent of the secretion of the thyroid gland, which enlarges to form a swelling in the neck, or *goitre*, if the supply is inadequate. Goitres occur in isolated continental areas far from the sea, and are prevented by adding potassium iodide to table salt.

Fluorine is necessary for hardening of the enamel of the teeth; an excess causes brown mottling of the teeth and increased density of the bones. Where the fluoride content of drinking water is low, dental caries is commoner and can be countered by adding fluoride to the water supply.

Copper may have some importance in the formation of haemoglobin. *Cobalt* forms part of the molecule of the B_{12} vitamin; and *zinc* is needed in some enzyme systems and for the healing of wounds. *Manganese* is also important to certain enzyme systems.

Nutrients

Complex though the constituents of living matter and foodstuffs may appear, they can be grouped into three classes: fats, carbohydrates and proteins.

Fats

The *fats* occur in the adipose tissue beneath the skin and elsewhere, and as globules in the cells everywhere, particularly in the liver. They are essentially combinations of glycerol (or glycerine) with three fatty acids – *oleic, palmitic* and *stearic*. The relative proportions of these acids determine the solidity and melting point of each fat, e.g. the fluidity of human fat compared with solid mutton fat at body temperatures. Oleic is the most fluid; palmitic and stearic being relatively hard. Fats are insoluble in water, dissolve easily in ether and alcohol, can be emulsified in water as fine dispersed droplets, and broken down during digestion into their constituent fatty acids, prior to rearrangement by the tissues.

Fats provide the largest source of energy next to carbohydrates, and in concentrated form, so they are valuable for manual workers and in cold climates. They are vehicles for the fat-soluble vitamins A, D, E and K. A distinction is made between certain essential fatty acids known as *polyunsaturated*, which have to be ingested as they cannot be synthesised, and the *saturated* animal triglyceride fats solid at body temperatures. The unsaturated fatty acids are abundant in the fluid vegetable and fish oils. Fats are oxidised to carbon dioxide and water, though, if the diet is deficient in carbohydrates, the breakdown may be incomplete and intermediate metabolic products appear in the urine – the condition known as *ketosis* because of the presence of ketones.

The grease of the skin, the oil of the hair and the wax of the ear are more complex fats known as *sterols*; and even more complicated combinations with phosphorus are found in nervous tissue and the bounding membranes of cells.

Carbohydrates
These may yield half the total calories in civilised communities, but some peoples are healthy with virtually none, so they are not essential, if only because the body can form glucose from fats and protein. They are the immediate products in plants of the synthesis of water and carbon dioxide in photosynthesis, and important in animals as a ready source of energy. They are compounds of carbon, hydrogen and oxygen, and their essential saccharide unit, CH_2O, is used as a building brick to form compounds of increasing complexity.

The *monosaccharides* include glucose, the sugar of fruits – also formed in the body by the digestion of cane sugar and starch – and fructose.

The *disaccharides* include cane sugar, the maltose of fermenting grain and barley, and the lactose of milk.

The *polysaccharides* are complex substances of high molecular weight. They include *starch*, found as grains within the plant cells and converted by digestion into glucose. *Glycogen*, or animal starch, is very important. It is stored as granules in all the tissues, particularly in the liver and muscles. The liver glycogen is the great energy reserve of the body, mobilised when required by conversion into glucose. There is also the rough *cellulose* of plants, not digestible to any extent by man.

Proteins

The *proteins* are the most important constituents of protoplasm. They occur in quantity in lean meat, cheese and the vegetable pulses, and are extremely complex unions of carbon, hydrogen, oxygen and nitrogen, and usually also of sulphur. They are colloids, soluble in water and coagulating on heating. The essential unit of the protein molecule is the *amino-acid*, and though there is an enormous variety of proteins, the only chemical difference lies in the particular amino-acids contained in the molecule, their properties and the arrangement of their peptide linkages. This is important because the more essential amino-acids are only found in the 'first-class proteins', such as meat.

The protein of each animal species is quite different. Together with the reaction a foreign protein excites in the living tissues of man, this causes the phenomena of *immunisation* and *allergy*. Protein is, in fact, the basic chemical unit of protoplasm and different tissues possess characteristic proteins: the red haemoglobin of blood, the slimy mucoprotein of mucus, the casein of milk, egg albumen, muscle protein and the complex nucleoproteins linked with phosphorus in cell nuclei. Proteins can function both as weak acids and as weak alkalis, and thus maintain a buffering neutrality in tissue fluids.

The proteins are not completely broken down in the tissues, and much of their nitrogen content is lost in the urine as urea, uric acid, and so on. Only 70% of the energy theoretically available in protein is actually utilised by the tissues, whereas with fats and carbohydrates breakdown is total, and the energy yield is exactly what would have been obtained by combustion in the laboratory. Nitrogen balance is between daily intake and daily loss in the urine. The balance is negative in starvation and wasting diseases, and after major injury or operation. The balance is positive – i.e. there is nitrogen retention – in growing children and convalescent patients on an adequate diet.

Within limits, fats and carbohydrates are interchangeable as sources of energy, but the diet *must* contain protein to supply nitrogen and certain essential amino-acids that the body cannot synthesise. In Western countries, about 10% of the energy intake is provided by protein, 40% by fat, and 50% by carbohydrate.

Enzymes

The complex changes of chemical breakdown and resynthesis are performed in the body much more rapidly than in the laboratory. Starch is completely converted into maltose by the saliva within a minute, though it takes the chemist several hours' boiling to achieve.

This facilitation of chemical changes is due to *enzymes*. We are familiar in inorganic chemistry with *catalysts* which are only needed in minute amounts to accelerate reactions in which they are not themselves consumed. Enzymes are the catalysts of the organic reactions of the body. They are complex colloidal materials, preferring a definite degree of acidity or alkalinity for optimum performance – e.g. the pepsin of the acid gastric juice, and the trypsin of alkaline pancreatic secretion. They facilitate breakdown, synthesis and oxidation processes, but especially the process of *hydrolysis*, when water is added to a molecule of a substance, often as a preliminary to its disintegration.

Enzymes are specific, each acting only on a particular material, known as its *substrate*, e.g. the amylase of saliva acts only on starch; and the digestive juices contain different enzymes designed to act on the various constituents of the food – lipases for fats, proteases for proteins, carbohydrases for sugars and starches. These do not act on other substances. Most enzyme reactions are reversible, and fall into two main groups: the processes of digestion occurring in the bowel from the outpouring of digestive juices; and the more intimate reactions, such as oxidation, occurring in the individual cells.

Most metabolic reactions occur only in the presence of an appropriate enzyme, and the rate of reaction is related not merely to the concentration of substrate and of enzyme, local pH and temperature, but also to the presence of certain *co-enzymes* and *activators*, which can be metallic ions, such as zinc or magnesium, or organic molecules. As proteins, enzymes can be *denatured* by heat, excessive acidity or alkalinity and certain metal ions. Many have been obtained in the pure or crystalline state. They are classified according to their effects as follows:

1 *Oxidoreductases* in oxidation/reduction;
2 *Transferases* in transfer of chemical groupings from one molecule to another;
3 *Hydroxylases* in hydrolysis of compounds into simpler molecules;
4 *Lyases* in splitting;
5 *Ligases* in joining;
6 *Isomerases* in rearrangement of molecules to form an isomer.

Metabolism

Physiologically, the life of man can be regarded as a continual production of energy by oxidation (burning) of food and the expenditure of this energy in (*a*) maintaining a body temperature above that of his surroundings, and (*b*) movement.

The balance sheet for this energy intake and output is exact; food burnt in the body liberates the same energy as when oxidised in the laboratory. However, the human arrangements are complicated because the food is not burnt *as such*, but only *after* it has been digested and assimilated into the living tissues. Only a few substances like alcohol can be burnt directly without first becoming part of the living protoplasm.

Thus there is a recurrent cycle of activities, known as *metabolism*. Part of this is the process of building up or repair, of assimilation of food into the tissues – *anabolism*; part is the breaking down of these tissues with the liberation of energy and excretion of wastes – *katabolism*. Both are going on all the time, though their relative proportions vary. Anabolism preponderates during growth; katabolism during starvation, and senescence.

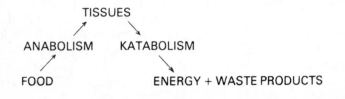

The wastes are excreted by the kidneys, bowel, lungs and skin. They consist mainly of water, carbon dioxide and nitrogenous breakdown products of protein, such as urea, which are mostly expelled in the urine.

During fasting, at complete rest, there is an essential minimum of energy output to maintain warmth and the movements of respiration and the heart. This minimum can be measured and is then known as the *basal metabolic rate*. Although of the same order of magnitude for everyone, it varies from person to person by virtue of their different sizes. This is because most of the energy is expended in making up for heat loss from the body surface, and is therefore directly proportional to the surface area. Basal metabolism is higher in men than in women, and higher still in young children. Each individual has his normal metabolic rate, predictable from height and weight, and this is increased in fever and excitement, and also by overactivity of the thyroid gland. In thyroid deficiency, the metabolism is considerably below normal.

From the basal rate we can estimate the calorie value of the food required to maintain the essential activities of the body; this is in the region of 2000 calories per day. Below this level the body begins to fall back on its reserves; the fat and glycogen stores are first consumed, and then the less essential tissues, such as the muscles and glands, digest themselves to feed the brain and heart – with eventual death when the latter fails.

3

The Skeleton

Cartilage

Cartilage, with bone, is a major component of the skeleton, the greater part of which is preformed in cartilage during intra-uterine life. It is found where rigidity and resilience are needed, i.e. at the joint surfaces, the front ends of the ribs, and the supporting framework of trachea and bronchi, nose and ears. The main types of cartilage are:

1   The smooth, clear *hyaline* cartilage of joint surfaces, with very low friction, produce a slippery surface.
2   Tough *fibrocartilage* containing white fibrous tissue, found in the intervertebral discs and in the plates or menisci that project into certain joints (e.g. the semilunar cartilages of the knee).
3   *Elastic* cartilage in flapping structures like the ear and epiglottis.

Cartilage is a gristly tissue, largely without blood supply, and so cannot repair itself after injury, being replaced by fibrous scar tissue. It is a connective tissue whose matrix has become solidified as a firm gel bound together by fibrous proteins (collagen and elastin). Its organic content includes the complex molecule of chondromucoprotein, linked with the polysaccharide chondroitin sulphate. The chondroblast or cartilage cell is plump and rounded, and distributed throughout the cartilage substance in small groups. It is responsible for the formation of cartilage matrix, and is nourished by diffusion from the fluid of the neighbouring joint and, to some extent, from the underlying bone. Disuse leads towards the degeneration of articular cartilage, whereas activity aids the diffusion of nutrients.

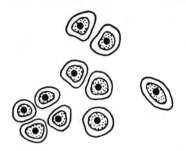

Fig. 3.1   Cartilage cells

Cartilage has the property of *calcification*, i.e. calcium salts are deposited in it, rendering it tough and opaque. Calcification is a normal process during growth, occurring as a preliminary to ossification of the cartilage precursors of the bones. But it is also an ageing process; in later life there may be fibrillary degeneration of joint cartilage, the wear-and-tear process of osteoarthritis.

Bone

Bone is also a specialised connective tissue of mechanical function, though it has an important connection with mineral metabolism, and houses the marrow in which new blood cells are formed.

It is the hardest tissue except for the teeth. It provides the framework of the body, protects the internal organs, gives attachments to the tendons and muscles, and constitutes the levers they move. Its structure combines strength with economy of material; its internal struts or *trabeculae* being disposed to meet maximum load. There is a considerable safety margin, e.g. the upper end of the femur can support a vertical load of a tonne and bears three times the body weight at every step. Bone is subject to compression, tension, twisting and bending strains, and withstands these by its strength and *elasticity*. In old age and some diseases this strength is impaired, and fractures occur.

Bone is a blend of (*a*) an organic fibrocellular matrix or *osteoid*, and (*b*) a *mineral* matrix, consisting of calcium phosphate and carbonate, magnesium and fluoride in crystalline form. The organic component can be removed by burning, leaving a brittle mineral skeleton; the inorganic matter can be dissolved by acids, leaving decalcified flexible bone. The osteoid consists of collagen fibres, plus some ground substance containing mucopolysaccharides and protein. The mineral matrix has a rigid crystalline structure, known as *hydroxyapatite*, represented as $Ca_{10}(PO_4)_6(OH)_2$, with variable amounts of other ions – magnesium, sodium, carbonate, citrate and fluoride. The collagen matrix acts as a site for the crystallisation of minerals from the dissolved calcium and phosphate of the tissue fluids under the influence of an enzyme, phosphatase, that concentrates phosphate ions and initiates crystal growth. In health the whole of the bone substance is mineralised, and unmineralised osteoid is only found where bone is being formed rapidly, e.g. at fracture sites. In certain diseases, such as rickets, mineralisation is defective, and the bone is soft and yielding. Sometimes, as in endemic fluorsis, mineralisation is excessive, and the bone is hard, but brittle. The salts of bone are in dynamic interchange with the ions of the body fluids, and this is the basis of the constant process of remoulding and replacement of bone substance. The calcium of blood is in equilibrium with that of bone, nearly 1 g being exchanged every day, a process influenced by the secretions of the thyroid and parathyroid glands, and by vitamin D and the local pH. Bone contains 99% of the total calcium of the body, 88% of the phosphate, 70% of the citrate, 80% of the carbonate and 50% of the magnesium. It is constantly subject to the influences of simultaneous formation and resorption. Both are very active during growth, with bone formation in the ascendant. During adult and early middle life the bone mass remains fairly constant in health, but in later life resorption leads and the amount of bone decreases so that it appears more translucent in an X-ray – *osteoporosis*. This process begins earlier in women, and accounts for the frequency of hip fractures in the aged.

Microscopically, the dense outer cortical bone of the shafts of long bones is arranged in layers, or lamellae, containing small clefts, or lacunae, occupied by the bone cells, or *osteoblasts*, where processes ramify around. The lamellae are arranged around a central Haversian canal containing blood vessels and nerves, and a bone is composed of a great many such Haversian systems, or *osteones*. This is modified near the surface of the shaft, where the lamellae run parallel and there are no canals. The osteoblasts are specialised connective tissue cells helping to form the organic

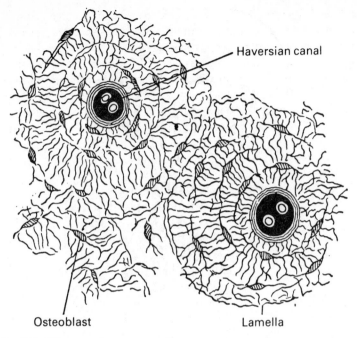

Haversian canal

Osteoblast   Lamella

Fig. 3.2 Microscopic structure of bone

matrix, and contain alkaline phosphatase, which assists their bone-forming activity. Wherever bone is being absorbed or remodelled, there is another type of cell, the *osteoclast*, a giant multinucleate cell that actively reabsorbs bone substance. The two types of cell work together to shape the internal architecture of bone in response to mechanical stress.

The *periosteum* is a tough ensheathing membrane surrounding bone, except where it is covered by articular cartilage. Its outer fibrous layer is mainly supporting; it carries part of the bone's blood supply and gives attachment to muscles and ligaments. From it fibres penetrate into the bone. In the deeper layer of the periosteum are osteoblasts, responsible for increase in girth of the bone during growth. During growth, the periosteum can easily be peeled off the underlying bone, especially if there is injury or infection. The attachment is much firmer in adult life.

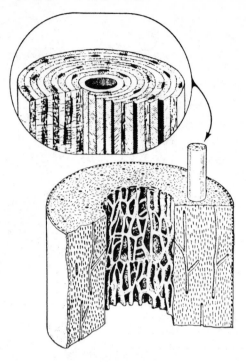

Fig. 3.3 Diagram of cortical bone showing the structure of an osteone. These rod-shaped units of bone structure are made up of series of lamellae (From *Electrical Effects in Bone* by C. Andrew L. Bassett. Copyright © 1965 by Scientific American Inc. All rights reserved)

Blood supply of bone

A typical long bone has four sets of vessels (Fig. 3.4). The *epiphyseal* vessels supply the epiphyses, or centres of ossification at the growing ends of the bone. Then there is a separate set of *metaphyseal* vessels, which are important in connection with growth. The shaft is supplied by one or more large *nutrient arteries*, which penetrate the cortex and enter the marrow cavity. Finally, the *periosteal* vessels nourish the outer layers of the shaft. If the nutrient artery is damaged, or if the periosteum is stripped off by infection, much of the shaft may die.

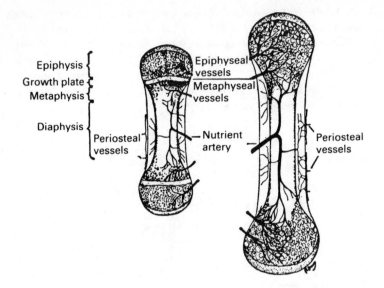

Fig. 3.4 Diagram showing the arrangement of blood vessels in bone before and after the growth period. Note that the growth plate isolates the epiphysis from the metaphyseal circulation during the growing period, but that they anastomose freely after growth has ceased. The periosteal vessels supply the osteogenic layer on the bone surface during the growth period, but penetrate the cortex and supplement the nutrient artery in adult life (From *A Companion to Medical Studies, Vol. 1*)

Types of bone
There are five main types of bone:

1   *Long bones* are those of the limbs, e.g. the humerus in the arm, the femur in the thigh. They have a shaft which is roughly cylindrical but sometimes polygonal or triangular in section, and two expanded ends, sometimes rounded off as a head or widened into condyles. The bone ends take part in the adjacent joints and are covered with smooth articular cartilage to facilitate movement; the two ends forming a joint are enclosed in a common joint capsule. The periosteum covering the shaft becomes continuous with the capsule at the end of the bone (Fig. 3.5 (b)). The bone surface is dotted with

numerous tiny apertures (*foramina*) for its blood vessels, with a larger nutrient foramen for the main artery near the midshaft.

A cross-section of a long bone shows the outer *cortex* of bony substance and the central *medullary cavity*. The cortex has an outer portion of dense compact bone surrounding a meshwork of spongy or cancellous bone; at the ends, where the medullary cavity stops, there is a solid mass of spongy bone (Fig. 3.5(a)).

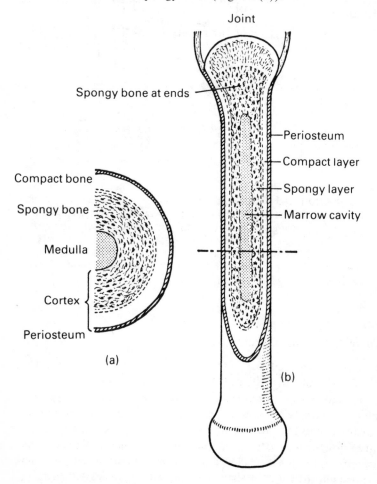

Fig. 3.5 (a) Cross-section of a long bone. (b) Longitudinal section of a long bone

Fatty *yellow marrow* occupies the medulla in the adult and the interstices of the spongy bone at the ends are filled with *red marrow*, responsible for the formation of red and white blood corpuscles. At birth, and in the early years, red marrow occupies the whole shaft, but retreats to the ends with growth. This can be reversed later if anaemia or haemorrhage make extra demands on blood formation; in chronic anaemia the shafts are occupied with red marrow working overtime.

In adult life active red marrow is found mainly in the flat bones, the skull, ribs, sternum (breast bone) and pelvis; also in the vertebrae of the spinal column.

2   *Short long bones* are simply long bones in miniature, found in the hands and feet.

3   *Short bones* are squat, cuboidal or irregular, composed entirely of spongy bone with a thin compact shell. They occur in the wrist (*carpus*), the corresponding part of the foot (*tarsus*), and the vertebrae.

4   *Flat bones* include the skull, vault, ribs and scapula (shoulder blade). They consist of two plates of compact bone sandwiching a thin spongy layer; in the skull these layers are called the inner and outer tables, with the diploë between. Certain bones of the skull are expanded by air-containing cavities replacing this spongy layer – the air sinuses.

5   *Sesamoid bones* are tiny, rounded masses found in certain tendons at points of friction; the largest is the patella at the knee (knee-cap).

Development and growth of bone

In the embryo of a few weeks, the central core of the primitive gelatinous connective tissue of the limb buds becomes transformed into an axial rod of cartilage. This rod is absorbed at the sites of the future joints, e.g. the elbow and knee, demarcating arm from forearm and thigh from leg, and split longitudinally in the distal segment as the forerunner of the paired radius and ulna, or tibia and fibula. Most of the skeleton is well formed in cartilage by the sixth week, and at the seventh week a centre of ossification develops in the midshaft of each long bone. Bone cells appear, the matrix is impregnated with calcium salts, and ossification spreads up and down the shaft until, at birth, the long bones are entirely ossified except for their cartilaginous ends.

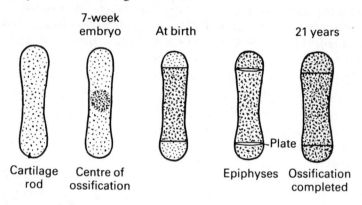

Fig. 3.6 Stages in the ossification of a long bone

Epiphyses

There now develops an arrangement, peculiar to mammals, allowing continuous growth in length during the years preceding maturity. In the first years of life separate secondary centres of ossification appear in the bone ends, which ossify completely except for a thin cartilage plate separating them from the shaft. The rounded end is called the *epiphysis* and the cartilaginous plate, the epiphyseal plate. This consists of longitudinal columns of cartilage cells reproducing themselves continuously on the shaft side of the plate and as continuously turned into new bone, which is pushed away into the shaft, enabling it to grow into length. Meanwhile, the epiphyseal plate retains its integrity, until the epiphysis fuses with the shaft by ossifying across the intervening cartilage – at the age of eighteen to twenty in men, sixteen to eighteen in women (see Fig. 3.6).

Growing ends

Although each end of a long bone is a growing end, one epiphysis appears earlier than the other and fuses later, and is mainly responsible for growth in length. In the arm these principal 'growing points' are at the shoulder end of the humerus, and the wrist end of radius and ulna; in the leg, at the knee end of all three bones. Complete destruction of an epiphysis causes cessation of growth and stunting; partial destruction causes a distortion of growth in which the limb turns away from the still active side.

A few bones, mainly the skull vault, mandible and clavicle, are not pre-formed in cartilage, but are ossified directly in primitive membranous tissue, and have no epiphyses.

Remodelling
The articular cartilage, or joint surface, which covers the free end of the epiphysis, persists throughout life. It is that portion of the original cartilage system that has not been substituted by bone formed in the secondary (epiphyseal) centre of ossification. The shaft of the bone grows in thickness by the formation of new bone within the periosteum. As bone is added externally, it is removed internally, so the shaft remains tubular, is strong without being massive and allows room for the bone marrow. This process of external deposition and internal reabsorption is called 'remodelling'. Remodelling also occurs at the metaphyses (the metaphysis is the 'neck' of the bone, just on the shaft side of the epiphyseal plate).

Growth

The foetus grows continuously, but its *rate* of growth varies considerably. It is slow during the first two months, reaches a peak at four to five months, and slows again in late pregnancy. The rate of growth falls off rapidly after birth and then more slowly until puberty, when a growth spurt occurs – the adolescent increment. All growth in height ceases at around eighteen years in boys and sixteen years in girls; this is because of epiphyseal fusion in the long bones. In old age height decreases, due to bending and degenerative changes in the spinal column. The changing *proportions* of the different components of the body are shown in Fig. 3.7. In early foetal life, the head is disproportionately large because of the precocious development of the brain; at maturity, the lower limbs make the major contribution. Before puberty, the legs grow faster than the trunk, and boys are generally taller than girls because they have a longer prepubertal period. Most of the adolescent increment occurs in the trunk.

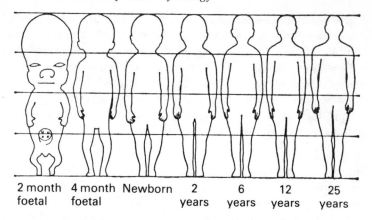

| 2 month foetal | 4 month foetal | Newborn | 2 years | 6 years | 12 years | 25 years |

Fig. 3.7 To show the changing proportions of the body during growth and development (From *A Companion to Medical Studies, Vol. 1*)

Factors controlling bone growth

As growth in height depends on growth in length of the long bones, which depends on growth in the epiphysial cartilage plates, growth rate and final stature are determined by the rate of proliferation of cartilage cells in the growth plate and its active duration. These are affected by a number of factors.

There is a strong *genetic* influence; identical twins are of similar height, and tall parents tend to have tall children. *Malnutrition* in childhood stunts growth. There is an intimate control of bone growth by certain *endocrine* glands. The growth hormone secreted by the anterior lobe of the *pituitary* gland promotes growth from birth to adolescence by its effect on the proliferation of cartilage cells. An excess of secretion in childhood so stimulates epiphysial growth as to cause *gigantism*, the victim reaching 2·13 or 2·44 m (7 or 8 feet) in height. Its action in the adult is less dramatic, but there is hypertrophy of the skeleton and soft tissues, especially of the hands, feet, face and tongue – the condition known as *acromegaly*. Deficiency of growth hormone in early childhood is one cause of dwarfism. The hormones of the *thyroid* gland are also important; in congenital deficiency (*cretinism*) there is stunting and idiocy, and the epiphyses are delayed in appearance and unfused in middle age. Sex hormones, secreted by the sex organs and the adrenal glands, cause the adolescent growth spurt.

Skeletal age

There is much variation in the calendar ages at which signs of maturity and sexual development appear – such features, for instance, as the onset of menstruation and the development of pubic hair. A boy or girl of chronological age of thirteen or fourteen may still be a child, or virtually an adult. True developmental age is best assessed from study of the dates of appearance of the secondary epiphysial centres of ossification in the long bones and the primary centres in short bones, such as those of the wrist. It is thus possible to arrive at a skeletal age that is a good index of maturity. Skeletal and calendar ages may differ by as much as two years.

The hard bones of the dried skeleton obscure their *plasticity* during life, for they respond to stresses, much as a tree responds to the wind and weather. The more work put on them, the more they hypertrophy, particularly in response to compression or muscular pull, and the internal arrangement of bone trabeculae is a mechanical solution to the engineering problems imposed.

Joints and movements

Types of joint

The articulations between the bones vary as regards mobility in different parts of the skeleton and some are quite immobile. The main types are:

1 *Fibrous*. As a zigzag *suture* between the skull bones, whose interlocking irregularities are joined by a thin fibrous strand, often obliterated by ossification in old age; as a *syndesmosis*, which allows some stretching and twisting, as the distal tibiofibular joint; or as a broad sheet of *interosseous membrane* between paired bones – radius and ulna, tibia and fibula.

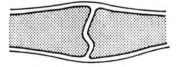

Fig. 3.8 Suture

2 *Cartilaginous*. Here the cartilage coverings of the bone ends are united by a plate of thick fibrocartilage, e.g. the cushion between

the two halves of the pelvis in front at the pubic symphysis, or intervening as an intervertebral disc between the vertebrae. Movement of such a joint is limited, but the total of movements between adjacent vertebrae may be considerable.

3   *Synovial joints* allow free movement. The bone ends are capped with smooth hyaline *articular cartilage*, and the joint cavity is enclosed within the sleeve formed by the fibrous joint *capsule*, stretching from one bone to the other and continuous with the periosteum. *Ligaments* of strong, fibrous tissue hold the bone ends together. Foremost of these is the main capsular ligament, and, in addition, there may be localised bands within the capsule and accessory ligaments traversing the joint or outside it altogether. Tendons arising within a joint function as extrinsic ligaments, e.g. the long head of the biceps at the shoulder joint. In some cases, complete or incomplete cartilage plates, or discs, project into the joint cavity as partitions; these are found at either end of the clavicle, between the lower jaw and the skull (temporomandibular

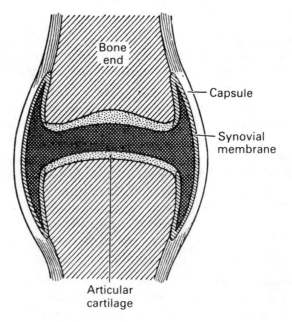

Fig. 3.9   A synovial joint

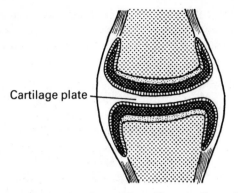

Cartilage plate

Fig. 3.10 Synovial joint partitioned by a cartilage plate

joint) and at the knee – the semilunar cartilages so often torn in footballers; also at the wrist at the inferior radio-ulnar joint.

Normally, the joint space is only potential, as the cartilage surfaces and soft tissues lie everywhere in contact under a negative pressure; it becomes a reality when the capsule is distended by fluid or if air is admitted. The capsule is lined by a smooth *synovial membrane*, which is reflected onto the bones but disappears at the periphery of the cartilage surfaces. This vascular, fatty membrane, with its folds and fringes, packs the joint and secretes the lubricant *synovial fluid*. This is a yellowish viscid fluid secreted from the blood, which also nourishes the articular cartilage; the white cells it contains scavenge the debris from the moving surfaces. Where the bones do not fit together accurately, there may be *fatty pads* between capsule and synovial membrane, which bulge into the dead spaces and cushion movement. Any structure traversing the joint, such as a ligament or tendon, carries an investing sheath of synovial membrane.

Stability
The integrity and coaptation of joints are maintained by a number of factors, the most important being the strength of the ligaments and the tone of surrounding muscles; muscle tone normally prevents a severe strain on the ligaments being more than momentary. Great mobility is essential at the shoulder, so the capsule is lax, and the large humeral head fits poorly into the shallow glenoid fossa of

the scapula; this inherent instability is overcome by having the tendons of the small rotator muscles blend with the capsule so that the anatomical deficiency is compensated by muscle guarding. This contrasts with the hip, where stability is important; the femoral head is deeply buried in the socket of the acetabulum and blending of the neighbouring muscles with the capsule is unnecessary. Minor supportive factors are interlocking of the bone ends, rarely very secure in itself, and the cohesion produced by atmospheric pressure.

The joint surfaces are least closely fitting during the main range of movement and most closely packed at the extremes of range. There is a *close-packed* position when the joint is arranged to face stresses and strains, and a *rest* position when it is not in use. Although the ligaments are important strengthening structures, they have little stretch and are not adapted to resist severe strain. Such movements are usually resisted by the surrounding muscles, which can remain tense throughout a resisted movement or pay out slack as their opponents contract in a manner impossible for the inelastic ligaments.

Varieties of synovial joints
Synovial joints are further subdivided by the shape of their bone ends, and the range of possible movements.

The simple *plane* joint between the small, flat adjacent surfaces of carpal and tarsal bones allows minor gliding motion only. The *saddle* joint is self-explanatory, with reciprocating surfaces, the best example being between the thumb metacarpal and the corresponding carpal bone. The *hinge* joint is a common mechanism, as in the elbow, ankle and fingers; here, a convex surface is gripped within a concavity, and strong collateral ligaments on each side allow motion round a transverse axis only. The capsule obviously needs to be lax in front and behind to allow full flexion and extension. The *pivot* joint has a cylindrical bone end rotating on its long axis like a hinge within its socket – e.g. the head of the radius at the superior (proximal) radio-ulnar joint at the elbow in rotation of the forearm.

The *ball-and-socket* joint of shoulder and hip has the greatest range, with one spheroidal and one concave surface moving on each other round an infinite number of axes. Its main movements (see Figs. 3.11 and 3.12) are:

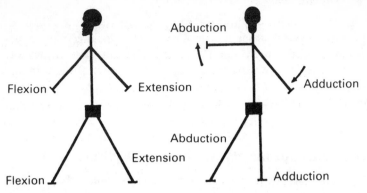

Fig. 3.11   Movements of the ball-and-socket joints of shoulder and hip

Flexion and *extension* in an anteroposterior plane; *abduction* and *adduction* in a mediolateral plane; *medial* and *lateral rotation* on the long axis of the limbs (not to be confused with pronation and supination of the forearm in the upper limb, where true rotation occurs only at the shoulder joint); and *circumduction*, in which the limb describes the boundary of a cone, in a combination of all the above.

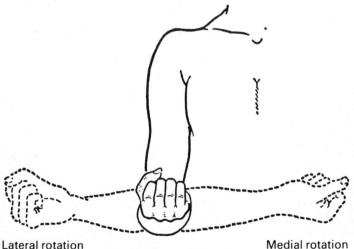

Lateral rotation   Medial rotation
Fig. 3.12   Rotation at the shoulder

The movements of joints are rarely as extensive as the shape of the bone ends would seem to allow, i.e. actual locking of the bones is only exceptionally a limiting factor. The soft parts are usually responsible for checking the range, as in contact of the arm and forearm in flexion of the elbow, and tension of the hamstring muscles at the back of the thigh in limiting flexion of the hip.

Axial and appendicular skeleton: limb girdles

The skeleton is divided into the central *axial* portion of the head and trunk, and the peripheral *appendicular* portion of the limb bones. The axial skeleton is made up of the skull, mandible (lower jaw), spine, sternum (breast bone) and twelve pairs of ribs. In addition, there are the small hyoid bone in the upper part of the neck beneath the floor of the mouth and three tiny ossicles in each middle-ear cavity.

Limb girdles

These are encircling arrangements of bones connecting the shoulder and hip regions to the axial skeleton, so as to provide a firm attachment for the corresponding limb while allowing a degree of mobility. The two girdles differ somewhat, as the arm does not bear weight and must allow the freest use of the hand; the shoulder girdle is correspondingly unstable and its main central connections are muscular rather than bony. These conditions are reversed at the hip, where stability is essential, and here the pelvic bones make up a complete bony ring, firmly attached to the spine behind (Fig. 3.15).

Both girdles are based on a circular model, but the shoulder girdle is very wide and open behind, where the scapulae are only attached to the spine by muscle so that they can move freely on the trunk. Each half of the shoulder girdle is made up of two bones: scapula and clavicle – the latter a slender bone joining scapula to sternum and bearing part of the burden of the hanging arm. Halfway round the girdle on each side is the shallow glenoid fossa of the scapula, accommodating the humeral head at the shoulder joint.

In the hip girdle, the bony ring is complete in front and behind in a tight circle. Each half is composed of a single pelvic (innominate)

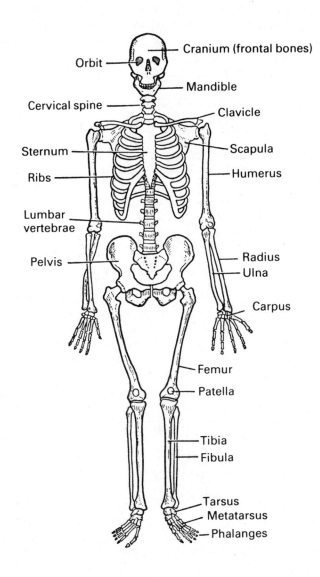

Fig. 3.13 The skeleton, anterior view

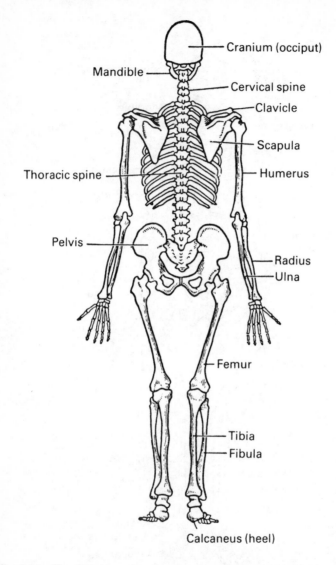

Cranium (occiput)

Mandible

Cervical spine

Clavicle

Scapula

Thoracic spine

Humerus

Pelvis

Radius

Ulna

Femur

Tibia

Fibula

Calcaneus (heel)

Fig. 3.14   The skeleton, posterior view

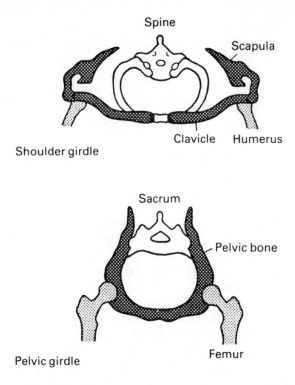

Fig. 3.15   The limb girdles

bone carrying a very deep acetabular socket for the head of the femur at the hip joint. Posteriorly, the two pelvic bones are firmly attached to the spine – here the sacrum – at the sacro-iliac joints. The girdle is integrated with the axial skeleton and does not permit any accessory movements except those at the hip.

4

Muscle

Apart from the amoeboid movements of some white blood cells and connective tissue cells, and the lashing movements of hair-like extensions of cells lining the air passages, all movement is the result of contraction in muscle cells. The heartbeat, the expulsion of urine, faeces or foetus, the rhythm of bowel movement, pupillary dilatation and contraction are all examples of such movements.

Muscular tissue is composed of reddish muscle fibres arranged in bundles. These fibres are highly specialised elongated cells characterised by their power of contraction on stimulation.

The muscles associated with the skeleton, and responsible for movements of the limbs and trunk, account for approximately half the body weight and contain half the body water. Their functioning is the chief factor governing energy production and expenditure, for even slow walking increases basal oxygen consumption five or six times.

Muscular tissues vary in the speed, strength and duration of their contractions. Some muscles act as the result of conscious effort, others function unconsciously. Some contract only if stimulated by nerve impulses, some have an inherent pattern of contraction and others respond to circulating hormones. These varying functions are associated with different types of muscle structure:

1 *Smooth* or plain muscle (unstriated, involuntary or visceral) is the most primitive. It forms the contractile coat of blood vessels and hollow internal organs like the bowel and bladder – structures that must work automatically, beyond conscious control, and regulated

Nucleus

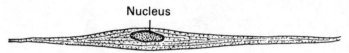

Fig. 4.1 A smooth muscle fibre

by the semi-independent autonomic or vegetative nervous system. The degree of control varies; some smooth muscle, such as that of the bladder, has an intrinsic rhythm which is only modified by nervous stimuli. All smooth muscle is stimulated to contract by being stretched. The typical smooth muscle cell is spindle-shaped, about 250 μm long and has a large oval nucleus at its middle.

2 *Striated* muscle (skeletal, somatic, voluntary) has more intricate fibres, which appear cross-striated under the microscope. They are found in the muscles attached to the skeleton and are under the conscious control of the central nervous system. These striped cells can be very long and may run for several centimetres from one attachment of the main muscle to the other. Each contains many hundreds of nuclei, placed to one side of the cell, and each cell receives near its midpoint the termination of a nerve fibre from the

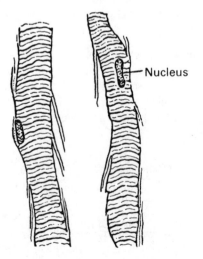

Fig. 4.2 Striped muscle fibres (After *Gray*)

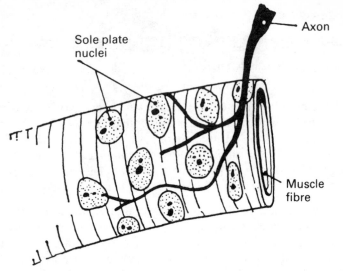

Fig. 4.3   Drawing of a motor end-plate from the rat. Note the terminal arborisation of the axon and the accumulation of nuclei in this region (From *A Companion to Medical Studies, Vol. 1*)

brain or spinal cord. At this junction, the muscle cell and the nerve fibre produce a complicated structure – the myoneural junction or motor end-plate.

3   *Cardiac* muscle, the wall of the heart, occupies an intermediate position. Its fibres are striated, but not under voluntary control, and the cells are branched so that the muscle forms a network. There are less than a dozen nuclei to each cell, situated near its centre. Each cell has its intrinsic rhythm and there is no clear demarcation between cells.

Fig. 4.4   Cardiac muscle

The muscle cell

The muscle cell has become specialised to convert chemical energy into contractile force, and is elongated along the axis of contraction. It is surrounded by an excitable membrane, the *sarcolemma*, while the cytoplasm is called *sarcoplasm*. Here the mitochondria are unusually large and numerous, there are many glycogen granules, and a special feature is the presence of contractile protein filaments, the *myofilaments*, which run the length of the cell and are only visible under the electron microscope. When the filaments group together and become visible to light microscopy, they are known as *myofibrils*.

Myofilaments are of two types: thick and thin. The thin filaments consist of a protein, actin, in fibrillary form, which can also exist in globular form. The thick filaments consist of another protein, myosin. Each molecule has a rounded head and a long tail; the tails of the molecules run together to form the thick strands, and the heads project at the sides of the filaments. The two kinds of filaments lie side by side, and the contractile force is produced between opposing molecules, where actin globules and myosin heads come together like the rows of a zip-fastener. As both filaments are polarised in the orientation of their molecules, contraction takes place in one direction only, i.e. a muscle can actively contract, but cannot actively lengthen.

In *smooth muscle* the filaments run along the long axis of the cell;

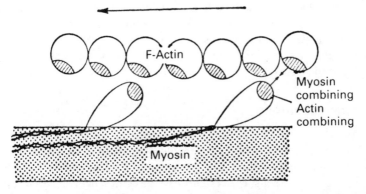

Fig. 4.5 The interaction of myosin and actin (From *A Companion to Medical Studies, Vol. 1*)

they are not regularly patterned, so no cross-striations are visible. There is no limit to the range of shortening, and smooth muscle, though it contracts slowly, can exert great tension for long periods – a *tonic* activity, under the influence of the autonomic nervous system and of circulating hormones, that is important in the functioning of the heart and blood vessels, bowel, bladder and air passages.

In *skeletal muscle* there is a regular arrangement in length of the two types of filament, producing the typical cross-striation associated with the grouping of filaments into myofibrils that allows rapid contraction. A muscle fibre can shorten by over half its relaxed length, which is the degree of contraction required of a skeletal muscle to produce a full range of movement at the joint it controls.

Additional smooth muscle cells can be produced in response to requirements throughout adult life – as in the uterus during pregnancy. This is not possible in skeletal muscle, where any increase in bulk from exercise or training is due to increase in size, or *hypertrophy*, of individual muscle cells. Muscles hypertrophy from the stimulation of sex hormones at puberty, especially in males. They *atrophy*, or waste, with disuse; interruption of the motor nerve supply produces severe and rapid wasting.

A general distinction can be made between the striped muscles attached to the skeleton and concerned with voluntary movements, controlled by the will through nervous messages from the brain, and the smooth muscle forming the walls of hollow internal organs, and of the blood vessels. The latter function without conscious attention, and are under the control of the semi-independent autonomic nervous system. Sometimes these two set of muscles are called the *somatic* muscles of the body wall and the *visceral* musculature of the internal organs.

Skeletal muscle

Voluntary muscles differ from the involuntary in that they contract in response to environmental changes, a response mediated by the central nervous system. Fig. 4.6 shows how a stimulus to the skin is conveyed by a sensory nerve fibre to the central nervous system and there relayed to a motor nerve cell, whose fibre runs out to end in the muscle it stimulates to contract. This is the *reflex arc*, well

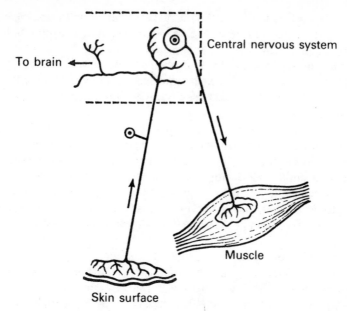

To brain ←

Central nervous system

Muscle

Skin surface

Fig. 4.6 A simple reflex arc. The stimulus conveyed from the skin is relayed through a cell of the central nervous system to elicit an automatic contraction of the appropriate muscle. At the same time information is passed to the brain for conscious control of the proceedings (Adapted from Starling's *Principles of Human Physiology*)

established in quite simple animals, and secures the immediate, unthinking performance of protective actions such as blinking. In man, reflex actions have come increasingly under the control of the will, i.e. the message is relayed up the spinal cord to the brain, which may modify the primitive response. The situation is like that of a ship being steered by a gyroscopic compass responding reflexly to any change in course, but providing information for the captain, who may intervene at any time.

Every skeletal muscle has motor nerve fibres arriving from the central nervous system, but it is also provided with sensory fibres carrying information as to the degree of tension in the muscle. This is important, as muscles act in groups and require central coordination based on knowledge of their behaviour.

Individual muscle cells either run the length of the main muscle or end at intermediate points. In the first case, all the end-plates lie in a zone near the centre; in the second case, they are distributed throughout the muscle.

Each muscle cell has one end-plate associated with the termination of a nerve fibre, and each motor nerve cell in the nucleus of a cranial or spinal motor nerve sends out one prolonged axon, or axis cylinder, within that nerve, which branches in the muscle to supply a number of cells. The nerve cell and its axon exhibit an 'all-or-none' characteristic: they transmit either an impulse of a particular magnitude or nothing at all. The arrival of an impulse at a motor end-plate causes a similar 'all-or-none' depolarisation of the related muscle cells, i.e. an individual fibre contracts by some 50–60% of its relaxed length or not at all. However, not all the fibres of a muscle necessarily contract at the same time, depending on the work to be done.

The group of muscle cells supplied by a single neurone is known as a *motor unit*, and the number of muscle cells it contains is related to the delicacy of the movement performed by that muscle, as there are fewer cells per unit where fine movements are concerned. In the small muscles controlling the movements of the eyes and fingers, a single motor unit comprises perhaps only a dozen cells, whereas in the coarser muscle of the limbs and trunk there may be several hundred. An electrode placed over or in a muscle can detect the electrical activities of local motor units, and these can be amplified to produce audible evidence of 'firing' – machine-gun-like effect – or the visual record of an electromyogram.

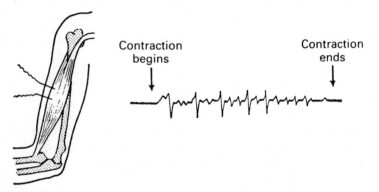

Contraction begins

Contraction ends

Fig. 4.7   Electromyogram

The transmission of the nervous impulse to the muscle cell at the end-plate replaces the electrical impulse by chemical stimulation. The impulse transmitted down the axon liberates acetylcholine at the end-plate, and this increases the permeability of the muscle membrane to sodium, potassium and other cations. An electrical potential is generated that spreads through the muscle fibre at several metres per second, and this can be recorded in the electromyogram. There is little electrical activity in a muscle at rest.

The situation differs in smooth muscle. A skeletal muscle is completely paralysed if its motor nerve is cut, but smooth muscle continues to function after interruption of its nerve supply, for the nerves modulate rather than initiate activity. Smooth muscle often has a double nerve supply; one group of fibres releases acetylcholine and the other, noradrenaline. The first is excitatory and the other inhibitory; smooth muscle also responds to circulating hormones.

Muscle contraction, tonus, posture

A muscle moves the joint or joints between its origin and insertion, the former means the fixed and the latter the moving attachment. This is clear enough when done against resistance. But even when a movement is assisted by gravity, as when the arm drops to the side, there is a controlled relaxation, which is really a continuous readjustment of the degree of contraction; i.e. 'relaxation' is an active process. Ordinary contraction, when origin and insertion are allowed to approximate freely and the muscle belly shortens, is *isotonic*; the tension remains constant throughout. If the attachments are kept apart by resistance, the muscle cannot shorten and its tension rises – an *isometric* contraction of constant length. Most actions are performed against some degree of resistance, so contractions are a blend of isometric and isotonic.

Living muscle is never completely relaxed, except under deep anaesthesia, but is always in slight contraction or *tonus*. Tonus is essential to posture, especially the erect posture, as it holds the body against the strain of its own weight. In the feet and legs, the ligaments would be subject to enormous stresses but for the guarding of the surrounding muscles. The spinal column is not a straight rod but a series of curves, whose general shape is maintained by the long spinal muscles forming the chords of these arcs. Postural reflexes in the erect position are devoted to preserving an upright, forward-looking head, and complex reciprocal action of all the muscle groups in calves, thighs, buttocks, spine and neck exists to this end. Hence the necessity for a *sensory* nerve supply from muscles, telegraphing the state of tension from moment to moment to the central nervous system.

The biochemistry of muscle contraction

The *immediate* source of the energy used in contraction is the splitting up of 'high-energy' adenosine triphosphate, which is involved in the actin-myosin reaction.

$$ATP + H_2O \rightleftharpoons ADP(adenosine\ diphosphate) + H_3PO_4 + 8\ kcal.$$

This ATP has to be resynthesized, so energy is required from other sources. Under *anaerobic* conditions, when sufficient oxygen is not available locally, as at the beginning of violent exercise, substances in the muscle fibres break down to provide energy for this resynthesis. The most important is glycogen, its degradation product being lactic acid. If this accumulates in the muscle, as it may during static contraction, fatigue and inhibition develop. Normally, the excess lactic acid diffuses into the circulation and is removed by oxidation; some is also taken up by the liver for reconversion to glycogen. Some of the lactic acid remaining in the muscle is restored to glycogen during the recovery period after contraction, but much is burnt away to carbon dioxide and water by oxygen under *aerobic* conditions. The glycogen stores depleted during heavy exercise may not be fully restored for several days.

In a steady state of exercise under *aerobic* conditions, fat and carbohydrate brought to the muscles by the bloodstream provide energy for resynthesis of ATP. The liver is an important source of glucose, but even this may become depleted, and in prolonged exertion hypoglycaemia may develop (a concentration of glucose in the blood below normal) and fat becomes more important as a fuel. In prolonged exertion the muscle stores of glycogen repair the carbohydrate deficiency and the lactate level in the blood rises.

An adequate supply of oxygen is essential to contracting muscle, which may use ten times as much as in the resting state. The presence of lactic acid in the bloodstream during exertion stimulates the respiratory and cardiac centres, both to oxidise circulating lactic acid and to deliver oxygen to the contracting muscles. The oxygen is needed, not so much to enable muscle to contract as to allow it to recover from the effects of contraction; a muscle stimulated in an inert gas is incapable of recovery.

The oxidative removal of lactic acid does not keep pace with accumulation at the beginning of continued effort, and its concentration in the blood rises. A main reason for breathlessness is to increase the oxygen supply, to get rid of excess lactic acid in the blood. When a steady state is achieved, the oxygen intake can deal with the acid as it is formed, and the individual has his 'second wind' and can continue comfortably. However, after a short period of very strenuous effort, one is out of breath for several minutes, as there is an accumulation of lactic acid in the blood still to be oxidised. The body has got into 'oxygen debt' by accepting more calls on its oxygen supply than it can meet at the time.

The situation is complex because the carbohydrate store in muscle is important as a short-term source of energy during the relatively anaerobic conditions at the beginning of exercise, before the local and general circulation have had time to adapt, whereas the main source of energy during sustained activity is the free fatty acids of the blood. In a short burst, such as a hundred metres' sprint, most of the energy comes from the breakdown of muscle glycogen and a large oxygen debt is incurred, i.e. the circulation has not provided oxygen and nutrients locally just when they were needed. Increase of cardiac output, the redistribution of blood to the muscles and increased intake of oxygen by respiration are not immediate. Also, muscles used rhythmically, as in running, only have an intermittent blood supply because the contractions themselves empty the blood vessels; again, there is partial oxygen deprivation and dependence on a diminishing local glycogen store. A step to overcoming this difficulty is the presence in muscle of a pigment, myoglobin, akin to the haemoglobin of the blood, which can liberate oxygen locally by dissociation. This is well developed in the 'red' pectoral muscles of birds.

Oxygen uptake from respiration increases with the rate of exercise. The maximum uptake for young men is around 3·5 litres per minute, for young women 2–3 l/min, and for trained athletes 6 l/min. The oxygen content of the atmosphere, the capacity of the lungs and the work of the heart are all limiting factors. The muscles may use the oxygen supplied more or less efficiently – the arm muscles being less efficient than the leg muscles. Other factors are the efficiency of the circulation and the haemoglobin content of the blood. If less haemoglobin is available to transport oxygen, as in anaemia, or if the atmospheric oxygen is inadequate to saturate the haemoglobin, as at high altitudes, muscle efficiency will be impaired until there has been an adaptive increase in the number of red corpuscles.

Much *heat* is produced during muscular contraction, partly from the oxidative processes, partly from mechanical work. Only some 25% of our energy output can be translated into mechanical work; the rest produces heat. Most of the heat generated by exertion is dissipated by the evaporation of sweat. But body temperature rises during the first hour of exercise, and thereafter the raised temperature is maintained. Bodily mechanics are more efficient at higher temperatures; hence the importance of warming-up for athletes.

Smooth muscle

Smooth muscle differs considerably from voluntary muscle in inherent properties and nervous control. It has inherent rhythmic contractions, and is under the control of a different nervous system. The motor nerves to voluntary muscle from the central nervous system have only to make it contract. The automatic fibres to smooth muscle are sympathetic and parasympathetic, stimulating it to relax and contract respectively. These actions are reproduced by certain chemical agents: adrenaline, which mimics sympathetic action, and acetylcholine, which mimics the parasympathetic. Smooth muscle is much more sluggish than voluntary muscle. It is particularly sensitive to stretching, which is important in hollow organs like the bowel and bladder, made of smooth muscle that responds to distension or relaxation. A *slowly* increasing distension, as when the bladder fills with urine, causes the muscular coat to relax proportionately so that the pressure of the fluid does not rise until the last stage of distension; then the rising pressure stimulates rhythmic waves of contraction, which end in mass expulsion of the contents of the organ. *Rapid* distension of a hollow organ provokes an immediate response.

The entrance and exits to hollow organs are controlled by circular bands of smooth muscle, or *sphincters*, e.g. the cardiac and pyloric sphincters at each end of the stomach; these normally contract until relaxation is necessary to allow the organ to fill or empty. The nervous and chemical control of a sphincter must be exactly the converse of that of the organ as a whole, since it must contract while the organ relaxes, and vice versa. So the parasympathetic and sympathetic nerves, and the corresponding chemical agents, have exactly the opposite actions on the muscular wall of an organ and the sphincter muscle that guards its exit.

Gross anatomy of muscles

The voluntary or skeletal muscle, the flesh of the body, makes up 40% of its weight and is divided into the *axial* muscles connected with trunk, head and neck, and the *appendicular* muscles of the limbs. The separate fibres are bound in bundles by connective tissue, and these in larger bundles to compose the individual muscle, with its sheath of deep fascia.

Attachments

Muscles are usually attached to bone or cartilage, sometimes to ligaments or skin, either directly by muscle fibres or by an intervening tendon, a sinew or 'leader', a cord-like structure of white fibrous tissue. Most muscles have a tendon at one or both ends; flat muscles have sheet-like tendon expansions or *aponeuroses*, e.g. those of the flank which extend from lower ribs to pelvis. When a muscle contracts one attachment remains fixed – the *origin* – and the other – the *insertion* – is moved towards it. Generally, in the limbs the origin is proximal and the insertion distal, while in the trunk the origin is medial and the insertion lateral. But origin and insertion are sometimes interchangeable. Thus the pectoralis major, inserted into the humerus, normally approximates the arm to the side of the chest. But if the arm is fixed by gripping a table, the same muscle pulls on the ribs at its origin, expanding the thoracic cavity as an accessory muscle of respiration, as in asthmatic attacks.

Attachments are usually just distal to the joint moved (see p. 101), e.g. the biceps into the upper radius just below the elbow; this is inefficient mechanically but allows speed of action. A muscle may have two or even three origins (*cf*. the biceps and triceps in the arm), but its insertion is nearly always single. Tendinous attachments to bone produce a ridged elevation, while direct muscular insertion leaves the bone smooth. Occasionally, muscles are attached to each other, usually between a pair on opposite sides of the midline; thus the mylohoids under the chin criss-cross at a line of junction, while the aponeurotic fibres of the abdominal muscles interlace in herring-bone fashion at the midline of the abdominal wall. A few muscles have two separate bellies connected by an intermediate tendon, to redirect the pull by tethering to a neighbouring bone. Some skeletal muscles act on soft tissues, e.g. those that move the eyeball and the soft palate.

The *form* of a muscle depends on the arrangement of its fibres. When these have a direct pull, the muscle may be fusiform (with a tapering belly), strap-like, quadrilateral or triangular (Fig. 4.8). More power is obtainable when the fibres are inserted into tendinous prolongations within the muscle substance, to give a concentration of short efficient fibres, as in the unipennate, bipennate and multipennate arrangements shown.

A strap-like muscle may be segmented transversely by tendinous intersections, e.g. the rectus abdominis of the abdominal wall (Fig. 8.7). In the hand and foot, where bulky muscles would obstruct movement, the bellies of the muscles moving the fingers and toes are contained in the forearm or leg, and continue as slender tendons into the extremities.

Muscles are named mainly after their actions, e.g. the *flexor pollicis longus*, the long flexor of the thumb; occasionally by shape, e.g. *pronator quadratus*, the square muscle pronating the forearm; or by some feature, e.g. the *quadriceps* of the thigh with four heads of origin.

Muscle groups and movements

Muscles are usually grouped with the same nerve supply, performing common or related actions. It is this movement that is represented in the brain. No individual muscle can be contracted by an act of will, nor can any muscle act alone. We try to lift our arm, *not* to contract our deltoid, though this is the main muscle concerned, and a whole host of associated muscles come into play. In relation to a movement, a muscle may fill one of the following roles:

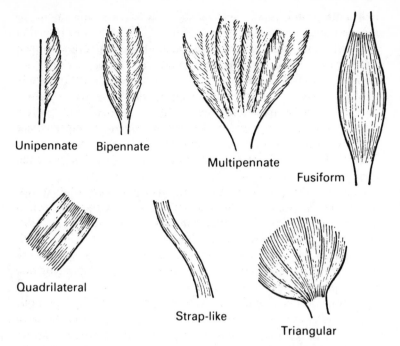

Unipennate   Bipennate

Multipennate

Fusiform

Quadrilateral

Strap-like

Triangular

Fig. 4.8   Variations in the form of muscles and the arrangement of their fibres

1   The *prime mover*, directly responsible for the movement;
2   The *antagonist*, which causes the reverse motion and has to relax reciprocally as the prime mover contracts, paying out slack to let the latter do its job;
3   The *fixators*, associated muscles steadying the base, or fulcrum, against which the prime mover acts;
4   The *synergists*, which control an intermediate joint so that the prime mover may act efficiently; thus the muscles closing the fingers can only produce a strong grip if the synergist muscles bend the wrist backwards.

In different movements the same muscle may play different parts.

In fine movements, such as writing, both prime movers and antagonists are in some degree of contraction throughout, a balance of opposing forces; in rapid forceful movements, such as striking a nail, both groups relax once the stroke is initiated and the movement continues by momentum.

Muscles may be excited to contract by heat, pinching, chemical irritation or an electric current; the last is of interest because the normal nervous stimulation of muscle is itself electrical. A constant galvanic current excites a single contraction at the beginning and end of its flow, at make and break; a faradic current does likewise, but the current is of much shorter duration and there is a minimum duration of flow below which there is no response. This *threshold value* is only a fraction of a second in warm-blooded animals. The response to electrical stimulation increases with the strength of current, and there is a minimum strength below which stimulation is ineffective. This increase of response is due to more fibres taking part in the contraction. But for each fibre, each contraction is an all-or-none effort; it contracts completely or not at all. After contraction there is a *refractory period* of a fraction of a second during which the muscle is inexcitable; this is a period of recovery and preparation for the next contraction.

If a muscle is repeatedly stimulated, it does not contract indefinitely but becomes *fatigued*; it does less work and eventually ceases to respond. This is due to using-up of all the food substances providing energy for contraction, e.g. glycogen, and the accumulation of wastes such as lactic acid. In combating fatigue a plentiful supply of blood and oxygen is essential. But fatigue in man does not only relate to the muscle itself; it also involves the tiring of the controlling nerve cells and fibres.

The generalised muscular contraction after death in *rigor mortis* must not be confused with contraction in life; it is associated with the accumulation of lactic acid and an irreversible change in muscle proteins. But it is accelerated by previous fatigue, so its onset is very rapid in animals that have been hunted to death.

5

Standard Positions, Terms and References; Body Coverings and Body Systems

Positions, terms and references

For the purposes of anatomical description, the body is always considered to be in a conventional position. The body is imagined erect, with the arms by the side and the palms facing forwards. The front of the body and limbs is the *anterior* surface, the back is *posterior*, so that for any two structures, one may be described as anterior or posterior to the other, insofar as it is nearer the anterior or posterior surface. Anterior is sometimes also called *ventral*, and posterior *dorsal*.

The position of structures is also related to the median plane of the body (AA in Fig. 5.1). One point is *medial* to another which is further from the midline, the second point *lateral* to the first. Points nearer the head end are *superior* to those *inferior* ones nearer the feet; *cranial* is sometimes used for superior and *caudal* for inferior.

Internal and *external* are descriptive of the boundary walls of body cavities or hollow organs; thus the ribs have an external surface directed outwards and an internal surface facing the thoracic cavity. *Superficial* and *deep* refer to relative distances from the surface, e.g. the skin is superficial in comparison to the muscles.

Certain other terms are used in connection with the limbs. Here, anterior may become *palmar* (hand) or *plantar* (sole of the foot), with *dorsal* for the posterior surface, the back of the hand or the upper surface of the foot. In the limbs lateral and medial may be replaced by names derived from the paired bones of forearm and

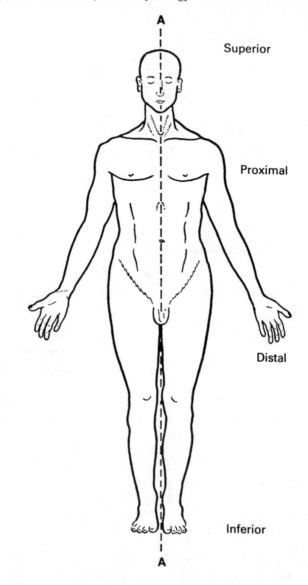

Fig. 5.1 The conventional view of the body from in front, showing its anterior surface (AA is the median plane)

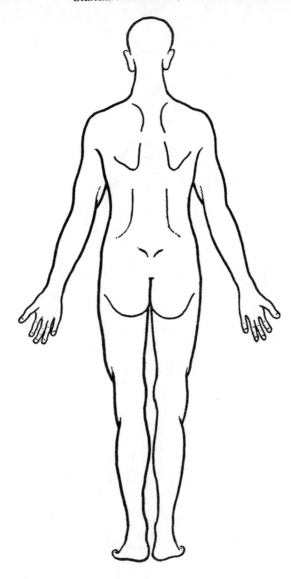

Fig. 5.2   The posterior surface of the body

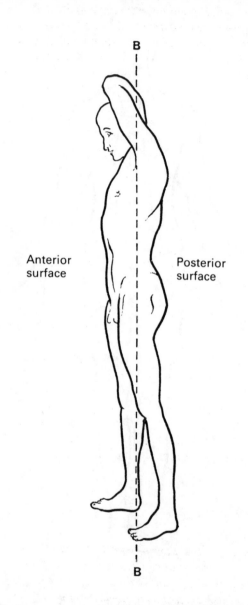

Anterior
surface

Posterior
surface

Fig. 5.3   The body in side view (BB is the coronal plane)

lower leg, i.e. *radial* and *ulnar* or *fibular* and *tibial*. Points nearer the shoulder or groin are *proximal* to those *distal* ones situated nearer the fingers or toes. *Peripheral* corresponds roughly to distal, but is usually employed for the outlying distribution of the branches of the circulatory and nervous systems.

It is often necessary to refer to sections through the body; these may be *horizontal* (transverse), *sagittal* (along or parallel to the median plane) or *coronal* (along or parallel to the coronal plane (BB in Fig. 5.3) at right angles to the median plane). All these terms must be mastered before proceeding; they are the cardinal points of the anatomical compass.

There is an agreed system of names for parts of the body, based on the original Latin terminology and revised at intervals, but it is not used in this book.

Symmetry and segmentation

The two halves of the body on each side of the median plane are similar, with corresponding right and left limbs, right and left kidney, and so on, i.e. many structures are *symmetrical*. However, as some internal organs are mainly or entirely one-sided, e.g. the liver and the spleen, and even the opposite limbs or the two halves of the brain are never exactly the same.

The human body repeats, in very modified form, the primitive arrangement of *segmentation* exemplified in the earthworm, which is made up of a number of identical segments, each containing the same organs. Integrated as it is, the human body retains features of this segmentation, seen in the arrangement of the vertebral column, in the series of paired ribs and in the segments of the spinal cord, each of which gives off a pair of spinal nerves to the appropriate body segment.

The body covering

Skin

The skin covers the body and is continuous with the linings of the internal canals at their orifices. It is elastic and mobile, except where bound down on the scalp, ears, palms and soles. On its surface open the hair follicles, sweat and sebaceous glands, and its brown melanin pigment is most developed on exposed sites, and at the genital and nipple areas. It contains the peripheral endings of sensory nerves, acts as an excretory agent through its glands and helps to regulate body temperature through water loss by evaporation. It is also protective, and is modified to form the hair and nails.

Microscopically, the skin has two main zones:

1   The superficial *epidermis*, thickest in the palms and soles, and permanently creased opposite the flexures of joints. It has an outer *horny layer* of dead or dying flattened cells, continually shed, and renewed by the growth of the *germinative layer* (see Fig. 5.4).

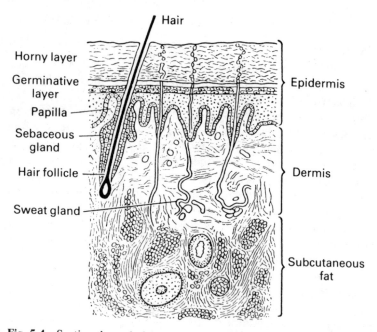

Fig. 5.4   Section through the skin to show its layers

Five layers may be recognised (see Fig. 5.5). Within the basal layer, a specialised group of cells produces melanin pigment. In dark-skinned individuals pigment is present in all the basal cells. The keratin of the horny layer is characteristic of skin and derived structures, such as hair and nails.

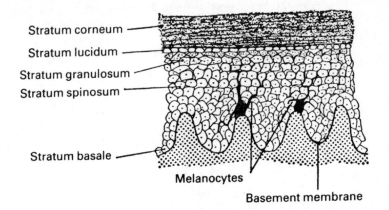

Stratum corneum
Stratum lucidum
Stratum granulosum
Stratum spinosum
Stratum basale
Melanocytes
Basement membrane

Fig. 5.5 The detailed structure of epidermis (From *A Companion to Medical Studies, Vol. 1*)

2 The deeper *dermis*, or true skin, is a layer of tough elastic and fibrous connective tissue between the epidermis and the subcutaneous fat. Unlike the epidermis, it is highly vascular and projects into the latter in little bays, or papillae, which contain the terminal capillary loops, and the end-bulbs of the sensory nerves for touch, pain, heat and cold. Although the hair follicles, sweat and sebaceous glands lie in the dermis, they developed ingrowth from the epidermis, and the hair shafts and ducts of the sweat glands traverse the latter to reach the surface. The sebaceous glands open alongside the hairs to ensure their lubrication.

The three to four million *sweat glands* occur over the whole skin surface and are most numerous on the palms and soles. They are simple, coiled, tubular glands, well supplied with blood vessels. *Apocrine* glands are modified sweat glands in the armpit, anogenital area and breast, related to sexual phenomena; they begin to function at puberty and produce the characteristic body odour. The individual *hair* consists of a long, dead, keratinised *shaft* and the basal growth segment or *bulb*, invaginated by the highly vascularised hair *papilla* of specialised tissue. Strands of involuntary muscle inserted into the follicles are responsible for erection of hairs, or 'goose flesh', in cold or fear. Hair growth is intermittent. After a growth phase, the lower part of the follicle degenerates, the hair shaft is loosened and shed as it is pushed out by the growth of a new hair.

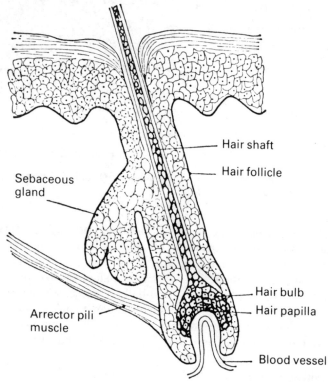

Hair shaft

Hair follicle

Sebaceous gland

Hair bulb

Hair papilla

Arrector pili muscle

Blood vessel

Fig. 5.6　Diagram of hair follicle and related sebaceous gland (From *A Companion to Medical Studies, Vol. 1*)

The *sebaceous glands* are found everywhere, except on the palms and soles. They are lobulated structures whose secretion, the *sebum*, lubricates the skin. Sebaceous secretion increases at puberty under the influence of the sex hormones.

The *nail* is a translucent keratin plate embedded in folds of skin. It is free at the tip, but elsewhere firmly attached to the underlying epidermis. The nail bed, from which growth occurs, is the epidermis underlying the basal fold. Nail growth is quicker at the fingers than the toes.

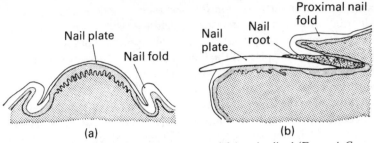

Fig. 5.7 Diagram of nail, (a) transverse, (b) longitudinal (From *A Companion to Medical Studies, Vol. 1*)

The protective function of the skin depends mainly on an intact epidermis. Substances may pass into the body directly via the epidermis or through the openings of the hair/sebaceous system. Penetration depends on the water and lipid solubility of the agent. The skin is almost impermeable to water, but fat-soluble substances such as alcohol enter because they dissolve the lipid of the cell walls; absorption is aided by the increase of skin temperature and blood supply, or if the skin is damaged. The intact skin is virtually impermeable to electrolytes and only slightly permeable to gases; in amphibia the skin is an important respiratory organ, but in man cutaneous respiration only constitutes 0·5% of total (lung) respiration.

Fasciae

Intervening between the skin and the muscles are two important layers of connective tissue, the superficial and the deep fascia.

The *superficial fascia*, the ordinary subcutaneous fat, is a continuous sheet over the whole body, fatty everywhere, save at the eyelids and male genitals. In one or two sites it contains muscle fibres, such as the muscles of facial expression, and the muscle which corrugates the scrotum. Fat is more developed in the abdomen, breast and buttocks, and thicker in women. It insulates the body, retaining warmth; and also contains the cutaneous nerves and vessels on their way to and from the skin.

Where the superficial fascia is abundant, as in the thigh, the skin moves freely over the deeper structures; where it is virtually absent, as over the nose and ear, the skin is firmly tacked down. The thicker deposit in women is responsible for their rounded contours; this distribution is a secondary sexual characteristic. In the male, fat tends to accumulate in later life in the upper abdominal wall. In certain sites – the pads of the fingers and toes, the heel-pad and the buttocks – the fat is honeycombed by fibrous strands; this is a cushioning device against pressure.

Internal to the superficial fascia is the *deep fascia*, a membranous sheet that covers and partitions the muscle groups, and lies in close relation to bones and ligaments. From its deep surface, other sheets, or *septa*, extend inwards between the muscle groups, and form sheaths for nerves and vessels and compartments for the viscera. This fascia varies greatly at different sites; it is virtually absent over the face, but extremely thick in the lower back.

The general arrangement of fasciae is seen in the cross-section of a limb (Fig. 6.25). This shows the skin, the superficial fascia and the deep fascia, enclosing the muscle masses in a restraining envelope and sending down between them septa that demarcate the various groups and reach the bone to blend with the periosteum. This is important for the return of blood and lymphatic fluid from the limbs. The heart pumps arterial blood into these compartments; the return of fluids towards the chest is due to the pumping action of muscular contraction within the non-yielding fascial envelope. As both veins and lymphatics have one-way valves, this pushes their fluid content towards the heart; if muscular contraction is eliminated by paralysis or immobilisation, tissue fluids accumulate in the limbs, the swelling known as *oedema*. In various situations the deep fascia forms restraining bands, or *retinacula*, to hold down tendons and prevent them from bowstringing when their muscles contract; *synovial sheaths* to facilitate the smooth gliding of tendons, as in the fingers and toes; and *bursae* – simple synovial sacs over points of friction, as between skin and bony points (knee-cap and elbow), and between tendon and bone.

Body wall and body cavities

The body wall, or *parietes*, encloses the great cavities, abdomen and thorax, which are also named after their lining membranes as the peritoneal and pleural spaces. The body wall consists of the skeleton, with its attached muscles and connective tissues, and the overlying skin and fat – the parietal or somatic structures.

The *cavities*, with their smooth serous linings, contain the internal, visceral or splanchnic organs: lungs and heart in the chest,

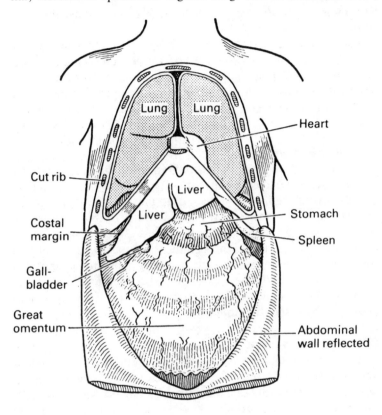

Fig. 5.8 The viscera exposed by removal of the anterior walls of abdomen and thorax

intestines and other organs in the abdomen. The viscera have developed in the embryo from the posterior body wall and still retain this attachment in the adult, the lungs by their roots, the intestine by its double-layered, fat-laden supporting mesentery.

The body systems

These are:

The skeletal system – the bones   ⎫
The joints or articulations   ⎬   the *locomotor system*
The muscles   ⎭
The respiratory system   ⎫
The digestive system   ⎬   the *visceral organs*
The urogenital system   ⎭
The vascular system – heart, blood vessels, lymphatics
The nervous system, and the sense organs

The digestive system

Food passing through the mouth enters an expanded cavity behind, the *pharynx*, which is common to the air passage at this level. It then travels down the gullet (*oesophagus*) to the *stomach*, thence to the *small intestine*, and then into the large intestine or *colon*. The waste matter in the food finally reaches the lowest part of the large bowel, or *rectum*, and is expelled through the short *anal canal*. At various points along the digestive tract, certain glands are situated and discharge their secretions. These are: the three pairs of *salivary glands* around the mouth; the *pancreas*, below the stomach; and the *liver*, which overlies the stomach.

The respiratory system

Air inhaled through the nose or mouth enters the pharynx behind and travels down through the air passage proper. The first part of this is the *larynx*, also the organ of voice; this leads on to the windpipe (*trachea*), which divides in the upper part of the chest into a right and left *bronchus* for the lungs. Each bronchus subdivides in the lung to form numerous branching *bronchioles* which end in clusters of tiny *air sacs*. It is in the walls of the latter that interchange occurs between the gases dissolved in the blood and those of the inhaled air.

Urinary system

The urine is secreted in the two *kidneys*, lying at the back of the abdominal cavity. From each kidney a tube, the *ureter*, carries the

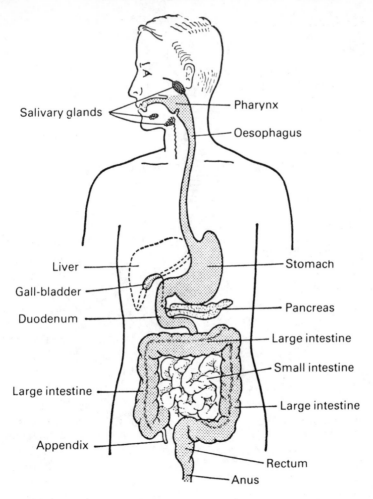

Salivary glands

Pharynx

Oesophagus

Liver

Gall-bladder

Duodenum

Stomach

Pancreas

Large intestine

Small intestine

Large intestine

Large intestine

Appendix

Rectum

Anus

urine down to the bladder in the pelvis, between the hip bones, and it is discharged from there into the *urethra*. In the female this is a short channel, soon opening externally; in the male it is a long pathway traversing the prostate gland and then running through the penis, which is also used for the reproductive act. The associated genital or sex organs will be considered later (p. 349).

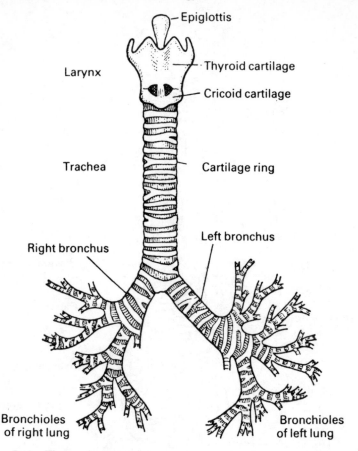

Fig. 5.10 The respiratory passages

Vascular system

The circulatory system is a closed circle, round which the blood is propelled by the contractions of the heart. Blood is driven into the arteries, thick elastic tubes that aid by their recoil the distribution of blood to all parts. The arteries divide into smaller branches in their course to the organs and limbs, and finally break up into a meshwork of fine capillaries, microscopic vessels that permeate every tissue of the body except the cornea of the eye and the outer layer of the skin.

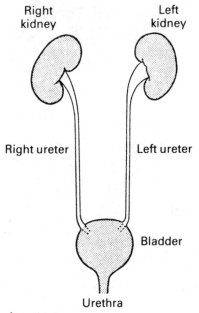

Right kidney

Left kidney

Right ureter

Left ureter

Bladder

Urethra

Fig. 5.11 The urinary system

The blood in the capillaries discharges oxygen and food materials to the tissue cells, and takes up carbon dioxide and wastes in return. The network reforms to form small veins, which become large venous trunks as they travel towards the heart. These are thin-walled, have no pulse and contain valves to prevent backward flow.

There are two separate circulations: a *systemic*, concerned with the body as a whole and driven by the left side of the heart, and a *pulmonary*, concerned with the passage of blood through the lungs and driven by the right side of the heart.

The right and left sides of the heart are shut off from one another; each with an upper chamber, or *atrium*,* receiving blood from the great veins, and a lower chamber or *ventricle* discharging blood into the great arteries. Stale venous blood from the body enters the right atrium, passes to the right ventricle and is expelled through the pulmonary artery to traverse the capillaries of the lungs. Here, it becomes aerated, receiving fresh oxygen in the air sacs and giving

*Sometimes called the auricle.

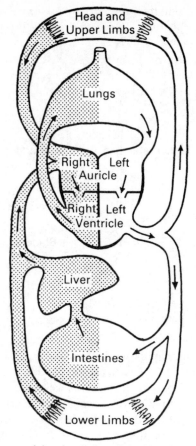

Fig. 5.12 A diagram of the circulation of the blood

up carbon dioxide to be breathed out. The fresh blood returns from the lungs in the pulmonary veins to the left atrium, then down to the left ventricle, and is discharged into the great artery of the body, the *aorta*, which supplies the head, trunk and limbs through its branches.

In the tissues the blood is rendered dark and venous, and is ultimately collected into great veins – the superior vena cava, draining the head and arms, and the inferior vena cava, draining the trunk and legs. Note that, whereas the arteries of the body contain bright red blood and the veins dark blood, the reverse is the case for the pulmonary arteries and veins, for the lungs reverse the chemical states of the blood.

A special arrangement of the abdominal vessels must be noted. Whereas the veins leaving most structures pass directly to the heart, those from the stomach and intestines enter another organ, the liver, where they break up into a second set of capillaries, so that the blood is filtered through the liver before reaching the heart. This ensures that the liver utilises and stores the food substances carried by the blood from the bowel; this is known as the *portal* circulation.

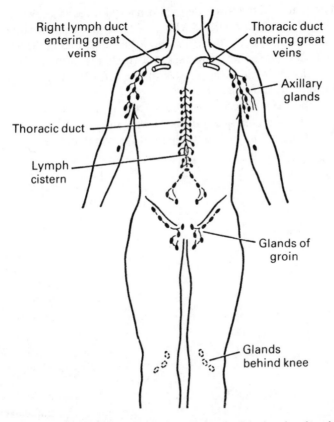

Fig. 5.13   The lymphatic system (The glands of the head and neck are not shown)

Lymphatic system

This is an accessory to the main vascular system. Not all the fluid portion of the blood that exudes into the tissues from the capillaries returns to these vessels, so there is an accumulation of tissue fluid. This excess is removed by a separate set of fine channels, the lymphatics, which begins as crevices between the cells and form a plexus draining the various organs.

These vessels pass up the limbs and trunk, and are interrupted at certain points by gland-stations or filters, which lie at the elbow and knee, armpit and groin, and in the trunk along the great blood vessels.

The lymphatics of the trunk join to form a wider vessel known as the *thoracic duct*, of matchstick thickness, which runs up in the chest to the left side of the neck; there it gathers the lymphatics of the left arm and the left side of head and neck, and discharges into the great veins. On the right side, the vessels of the arm, head and neck discharge directly into the veins.

One of the main functions of this system is the absorption of digested fat via the lymphatics of the bowel, and the lymph glands deal with any infection brought to them by the lymphatics.

6

Regional Anatomy: the Arm

Bones of the upper limb

Shoulder

This is made up of the rounded *shoulder-cap*, the prominence formed by the head of the humerus and the overhanging acromion process of the scapula. It also includes the *scapular* region behind, over the shoulder blade, the *pectoral* region or front of the upper part of the chest below the clavicle, and the *axilla* or armpit between the two. The shoulder girdle is made up of scapula and clavicle, articulating at the acromioclavicular joint; and the medial end of the clavicle articulates with the sternum at the sternoclavicular joint.

The *scapula* consists of the main *body* or blade, a thin, triangular plate carrying certain elevations or processes. The body has superior, medial (vertebral) and lateral (axillary) *borders*, with a superior and inferior *angle* at either end of the medial border. But where the corresponding lateral angle would be expected, at the junction of lateral and superior borders, there is the expanded mass of the *head* of the scapula, hollowed out on its lateral aspect to form the shallow *glenoid fossa*. The deep (anterior) surface of the bone is slightly concave and is applied to the backs of the upper ribs; the superficial (posterior) surface carries a prominent ridge, the *spinous process*, which runs up and out from its root on the medial border to end as a lozenge-shaped expansion, the *acromion process*, overlying the shoulder joint. Lastly, there is the stubby *coracoid process*, like a crook, arising from the superior border and overhanging the joint in front. Fig. 6.2 shows how the acromion and coracoid together form a protective arch over the joint.

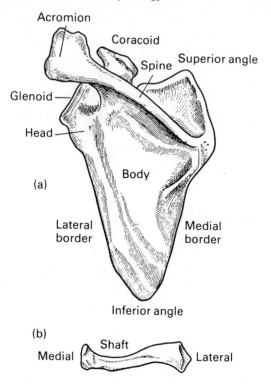

Fig. 6.1   (a) Left scapula, posterior aspect, (b) Left clavicle, from above

The *clavicle* is a slender, rod connecting the acromion to the upper portion of the sternum, the manubrium. It lies horizontally, forming the lower boundary of the neck at each side. Its ends are somewhat expanded and the shaft has an S-shaped curve.

Arm
The *humerus*, the bone of the upper arm, articulates with the scapula at the shoulder and with the forearm bones at the elbow; it has a shaft and two expanded ends. The upper or proximal end carries the rounded *head*, directed medially and upwards. On the lateral aspect, opposite the head, are two prominences, the *greater* and *lesser tuberosities*, giving attachment to the small rotator muscles which surround the joint. The greater tuberosity forms the

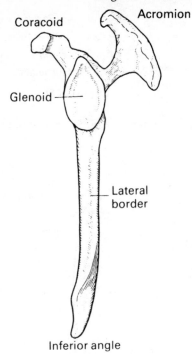

Coracoid

Acromion

Glenoid

Lateral border

Inferior angle

Fig. 6.2 Left scapula, lateral view

point of the shoulder beneath the overhanging acromion. The tuberosities are separated by the *bicipital groove*, which carries the tendon of the long head of the biceps from its origin within the joint, at the upper border of the glenoid fossa, on its passage into the arm. The true *anatomical neck* of the humerus is the narrow strip immediately encircling the head; the junction of the shaft proper with the whole mass of head and tuberosities is the *surgical neck*, a common site for fractures.

The *shaft* is cylindrical but the lower third is triangular in section. Halfway down, its outer aspect shows the rough *deltoid tubercle*, the insertion of the deltoid muscle which abducts the humerus from the side. Curving from the back of the bone, just distal to the tubercle, is the groove for the radial nerve as it passes to the front of the arm.

At the expanded lower end, the prominences above the elbow joint on each side are the medial and lateral *epicondyles*, the former

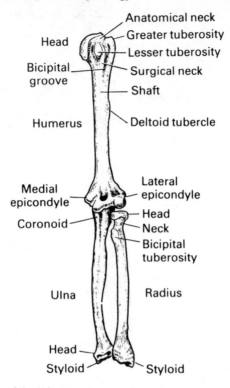

Fig. 6.3   Bones of the left upper limb, anterior aspect

is more developed as it is the origin of the strong flexor muscles of wrist and fingers. The polished, rounded projections from the lower end of the shaft – the lateral *capitulum* and the medial *trochlea* – are concerned in the elbow joint; there are hollow depressions just above them, the *radial* and *coronoid fossae* in front and the *olecranon fossa* posteriorly, which accommodate the bones of the forearm in flexion and extension.

Forearm

The *radius* and *ulna* are paired bones, articulating with each other at either end to form the superior (proximal) and inferior (distal) radio-ulnar joints; the lower end of the radius (but not of the ulna) forms the wrist joint with the carpal bones.

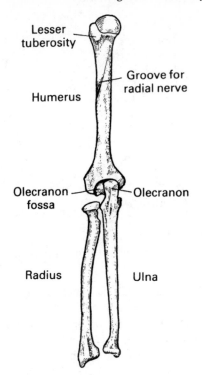

Lesser
tuberosity

Groove for
radial nerve

Humerus

Olecranon
fossa

Olecranon

Radius

Ulna

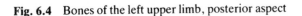

Fig. 6.4 Bones of the left upper limb, posterior aspect

The *ulna* is larger and lies medially; its shaft has a sharp border, felt under the skin throughout the back of the forearm. The upper end carries the olecranon process behind, the point of the elbow, which fits into the olecranon fossa of the humerus, and the coronoid process in front, corresponding to the coronoid fossa. These two processes are separated by the C-shaped trochlear notch, which embraces the trochlear process of the humerus; and just below this on the outer aspect of the ulna is a notch for the head of the radius. The head of the ulna is at its lower end, a knob on the back of the wrist, carrying the small, pointed styloid process.

The *radius* is shorter than the ulna, lying on the lateral side of the forearm; its larger end is distal and the small head proximal. The head is a smooth disc, hollowed-out above to receive the capitulum

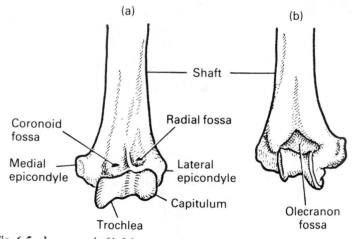

Fig. 6.5 Lower end of left humerus, (a) anterior, (b) posterior

of the humerus, and it fits within the ring formed by the radial notch of the ulna and the annular ligament of the elbow joint, rotating during the movements of pronation and supination (Fig. 6.6(a)). The neck is a constriction immediately below the head, and just below this on the inner side is the bicipital tuberosity for the

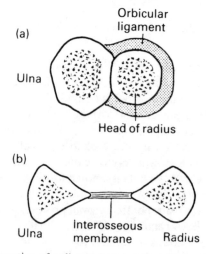

Fig. 6.6 Cross-section of radius and ulna, (a) at elbow, (b) at midforearm

insertion of the biceps. The broad lower end of the radius carries a styloid process which lies rather lower than the ulnar styloid; the back of the bone here is grooved by the extensor tendons of the wrist and fingers.

Both forearm bones have sharp, facing, interosseous borders, connected in life by the broad sheet of interosseous membrane, which runs the length of the forearm, and separates the anterior flexor and posterior extensor compartments (Fig. 6.6(b)).

Wrist and hand

The wrist or *carpus* is composed of eight carpal bones, small, irregular structures, arranged in two rows of four, a proximal and a distal. They are, from medial to lateral sides: proximal row – pisiform, triquetrum, lunate, navicular; distal row – hamate, capitate, multangulum minor, multangulum major. They are bound by strong interosseous ligaments so that the movement at any one joint between is small, but the composite range gives the wrist its flexibility.

The skeleton of the *hand* consists of five *metacarpal* bones; each is a short long bone, with a shaft and expanded ends. The proximal *base* of each articulates with a carpal bone at a carpometacarpal joint, and the rounded distal *head*, which forms the knuckle, makes up the metacarpophalangeal joint with the proximal phalanx of the corresponding finger.

The fingers contain three *phalanges*, articulating at the two (proximal and distal) interphalangeal joints. These phalanges – proximal, intermediate and terminal – are also short long bones; each has a proximal base and a distal end, formed by two condyles making a hinge joint with the base of the phalanx in front. At the end of the finger the terminal phalanx ends in a tuft which supports the nail. The phalanges are concave on their anterior (palmar) surfaces, because they form the floor of a tunnel roofed by fibrous tissue, through which the flexor tendons of the fingers glide in their synovial sheaths.

The *thumb* is specialised and differs from other digits. The metacarpal is short and slight, and is set freely away from the hand so that it can be opposed to the fingers. There are only two broad phalanges.

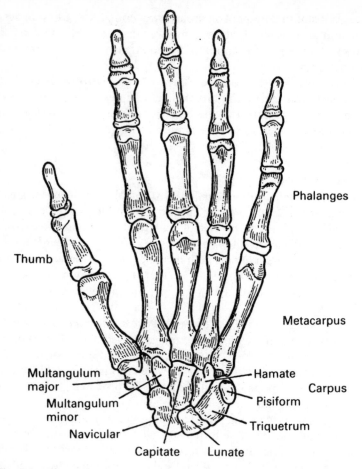

Fig. 6.7 Bones of left hand, anterior aspect

The joints of the arm

Clavicular joints

The *acromioclavicular* and *sternoclavicular* joints at either end of the clavicle are simple plane joints, allowing limited gliding. Each contains a projecting fibrocartilage disc or meniscus. They are important mainly to complement shoulder movement. The outer is easily dislocated by a fall on the arm, the clavicle riding up over the acromion.

Shoulder joint

This is designed for mobility at the expense of stability. The large rounded humeral head, about half a sphere, does not fit well into the shallow glenoid fossa. But the socket is deepened by a fibrocartilaginous rim, the *glenoid labrum*. The joint capsule is attached round the margins of the glenoid and the humeral head as a lax sleeve, hanging down in a fold below the joint; this laxity is necessary to allow the arm to be lifted from the side.

This instability is compensated in various ways. Several thickened bands in the substance of the capsule form accessory ligaments; and the tendon of the long head of the biceps originates within the joint from the upper pole of the glenoid fossa, traversing the cavity to emerge in the bicipital groove. The main support is the blending

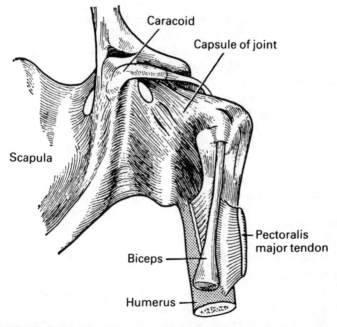

Fig. 6.8 Left shoulder joint, seen from in front (Note the loose capsule and the emerging tendon of the long head of biceps)

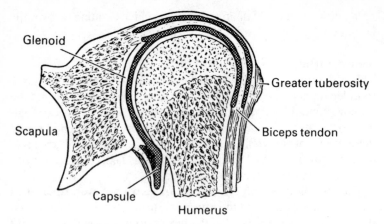

Glenoid

Greater tuberosity

Scapula

Biceps tendon

Capsule

Humerus

Fig. 6.9   The shoulder joint, longitudinal section (Note the dependent fold of capsule below, and the biceps tendon traversing the joint cavity)

with the capsule of the tendons of the four small rotator muscles arising from the scapula as they pass to their insertion on the tuberosities. These muscles are the supraspinatus, infraspinatus, teres minor and subscapularis. The muscles which move the joint also guard against displacement by tightening the capsule and holding the humeral head against the glenoid. Even so, dislocation is common, the head passing off the glenoid to lie under the caracoid process (Figs 6.22 and 6.23).

The movements of the shoulder are:

1   *Abduction* and *adduction*. In *abduction*, the arm is lifted away from the side in the coronal plane by the deltoid muscle; the reverse movement of *adduction* is effected by gravity, or actively by the pectoralis major muscle in front of the chest, and by the latissmus dorsi and teres major behind (Fig. 3.11).

2   *Flexion* and *extension*, which occur in the sagittal plane, through a combination of muscles (Fig. 3.11).

3   *Rotation*, produced by the small rotators, arising from the scapula (Fig. 3.12); it is lateral or medial.

4   *Circumduction* which combines the above movements, in a swinging motion of the outstretched arm.

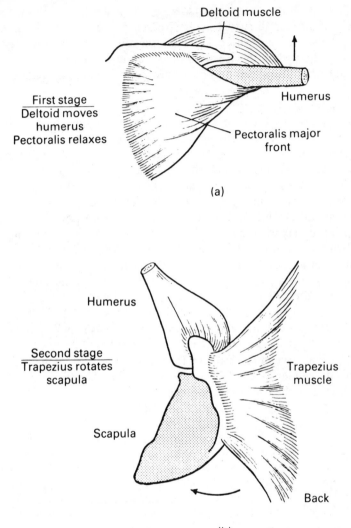

Fig. 6.10 Stages in abduction of the left shoulder joint, (a) anterior, (b) posterior. These stages are not as distinct as the figures suggest, for humerus and scapula both move from the outset, the former more at the beginning and the latter more at the end of movement

Abduction does not occur alone, but in association with movement of the scapula, which can be rotated on the trunk by the large trapezius muscle of the back. The arm can be abducted through 180°, but only half this is true shoulder motion, the rest is scapular rotation. Both movements occur simultaneously from the outset.

Elbow joint
This comprises three articulations: the true elbow hinge joint between the trochlear process of the humerus and the corresponding notch of the ulna; the shallow articulation of the radial head and the humeral capitulum; and, the proximal radio-ulnar joint.

As in all hinge joints, the collateral ligaments at the sides are very strong, and with the interlocking of humerus and ulna, allow only flexion and extension. The anterior and posterior portions of the capsule are lax so as to be easily stretched at the extremes of range. When the elbow is fully extended, arm and forearm are *not* in line but at an angle of 5–10°, the carrying angle – larger in women – which disappears as the joint is flexed. *Flexion* is produced by the biceps and brachialis muscles, the superficial and deep muscles of the anterior compartment of the arm; it is checked by the contact of the soft tissues of arm and forearm. *Extension* usually goes a little beyond the straight position, and is produced by the triceps at the back of the arm and checked by the fitting of the olecranon process into its humeral fossa. The triceps tendon blends with, and strengthens, the back of the capsule.

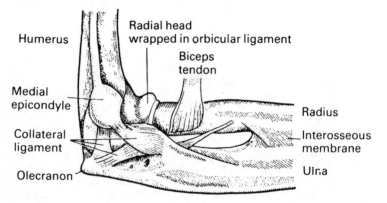

Humerus

Radial head
wrapped in orbicular ligament

Biceps
tendon

Medial
epicondyle

Radius

Collateral
ligament

Interosseous
membrane

Olecranon

Ulna

Fig. 6.11   Medial aspect of left elbow joint (After *Gray*)

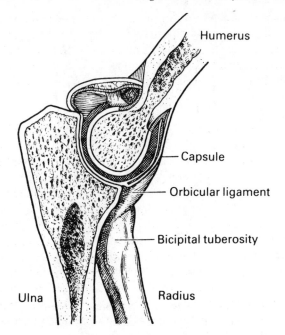

Fig. 6.12 Elbow joint, longitudinal section (After *Gray*)

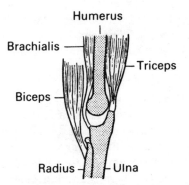

Fig. 6.13 Muscles responsible for flexion and extension of the elbow

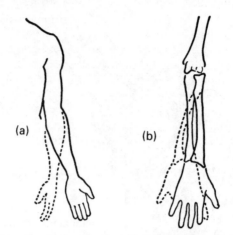

Fig. 6.14   Pronation and supination. Note in (a) the 'carrying angle' in the supine position, and in (b) the movement of the radius

The *proximal radio-ulnar joint* is formed by the round head of the radius rotating like a pivot in the embrace of the radial notch of the ulna and the annular ligament (Figs 6.6, 6.11).

Pronation and supination are the rotary movements of the forearm which twist the palm forwards and backwards. The ulna remains stationary while the radius rotates round it, moving at both radio-ulnar joints and along the axis of the connecting interosseous membrane. At the inferior radio-ulnar joint, the small ulnar head fits into a notch on the broad radius. In addition, the bones are connected here by a triangular cartilage plate with its apex fixed to the ulnar styloid process. This apex is the lower end of the axis of pronation-supination, the upper being the centre of the head of the radius.

Joints of wrist and hand
The *wrist joint* is formed between the proximal row of carpal bones (navicular, lunate, triquetrum) and the distal end of the radius, plus its cartilage plate. It is a modified hinge joint with strong collateral ligaments whose main movements are flexion (*palmarflexion*) and extension (*dorsiflexion*). There is also movement of the wrist and hand to either side, *abduction* and *adduction* – ulnar and radial

deviation – and the normal position for a strong grip is in slight adduction.

There are numerous small *carpal joints* between the individual bones of the carpus, fastened by short interosseous ligaments, the whole forming a complicated cavity continuous with the wrist-joint proximally and the *carpometacarpal joints* distally, where the metacarpals articulate with the distal row of carpal bones. Little motion occurs at the latter, except at the specialised saddle joint between the thumb metacarpal and the multangulum major. Here motion is very free and its components are:

1 *Adduction* to the side of the hand and *abduction* away from it, in the plane of the palm;
2 Movement in a plane at right angles to the palm, *palmar abduction* and *adduction*;
3 The characteristic human movement of *opposition*, in which the thumb is carried across the palm and its tip opposed to another digit. In this complex motion, the metacarpal rotates so that the thumbnail faces forwards instead of outwards (Fig. 6.15).

The *metacarpophalangeal joints* between the heads of the metacarpals and the bases of the phalanges are hinge joints, with strong collateral ligaments and a loose dorsal and palmar capsule. The

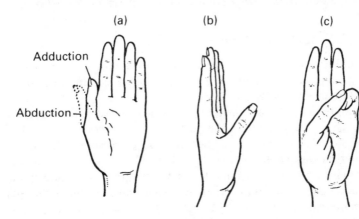

Fig. 6.15 Movements of the thumb

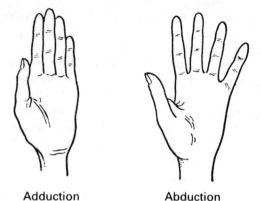

Adduction   Abduction

Fig. 6.16 Adduction and abduction of the fingers

main movements are flexion into the palm and extension backwards; usually there is 10–15° of hyperextension behind the plane of
the palm. In addition, there are side-to-side movements, known as
adduction and abduction, based on an axis through the middle
finger, abduction being away from this, and adduction towards
it. The middle finger is abducted whether it moves laterally or
medially.

The small *interphalangeal* joints allow flexion and extension only.

Surface anatomy of the arm

Some aspects of the surface anatomy will now be considered.
Inspecting and palpating the *anterior aspect* of the arm, the following features are noted:

At the *shoulder* the clavicle is seen under the skin; the acromion
process and upper end of the humerus give the rounded contour,
and the coracoid process can be felt by deep pressure in the hollow
under the outer end of the clavicle. The bulge of the deltoid muscle
is obvious, clothing the lateral aspect of the joint, and the pectoral
muscles appear when the arm is adducted against resistance. In the
arm there is the bulge of the biceps.

At the *elbow* the prominences of the epicondyles are visible and
palpable on either side, and the biceps belly fades into its tendon,
which is felt as it crosses the elbow hollow, the *antecubital fossa*.

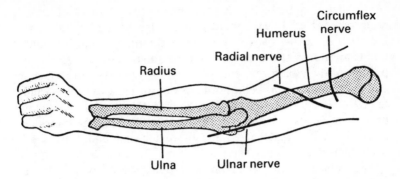

Fig. 6.17 Surface markings of important vessels and nerves of the left arm, posterior aspect

In the *forearm* the muscular bulge of the flexors of the wrist and fingers fades into their visible or palpable tendons just above the wrist, and on each side of the latter joint are felt the styloid processes of radius and ulna, the former at a slightly lower level.

In the *hand* the anterior aspect consists of the palm, with a muscular eminence on either side. The larger lateral *thenar* eminence, the ball of the thumb, is made up of the short muscles of that digit, and the shallower *hypothenar* eminence medially covers the muscles of the little finger. The three main palmar creases are fairly constant, and the line of the webs between the fingers as seen from the front is not opposite the metacarpophalangeal joints, i.e. the knuckles, but a good inch more distal.

Running down the *posterior aspect* of the limb, at the *shoulder* the

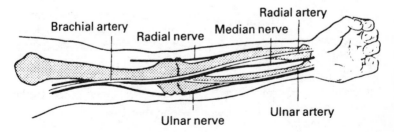

Fig. 6.18 Surface markings of important vessels and nerves of the left arm, anterior aspect

subcutaneous spine of the scapula ends in the overhanging acromion process, and again there is the bulge of the deltoid. The muscle mass at the back of the *arm* is the triceps, and the bony points are noted at the elbow, with the addition of the olecranon process of the ulna forming its point midway between the epicondyles. In the *forearm*, bellies of the extensor muscles of wrist and fingers can be seen or felt, and at the *wrist* the head of the ulna is prominent on the dorsum. On the back of the *hand* the knuckles mark the metacarpal heads, and the extensor tendons are seen when their muscles contract.

The surface markings of certain underlying nerves and vessels are indicated in Figs 6.17 and 6.18. The only superficial and easily felt nerves are the ulnar, in its groove behind the medial humeral epicondyle at the elbow, which can be rolled to give the familiar tingling, and the median at the front of the wrist. Both are vulnerable at these sites. The main arteries are deep except at the wrist, where the radial and ulnar vessels can be felt pulsating on either side; it is often possible to see the beating of the superficial arterial arch in the palm.

The subcutaneous structures of the arm

Fig. 6.19 shows the flayed arm, divested of skin and superficial fascia. It demonstrates how the deep fascia forms a continuous sheath for the deep structures, and a network of veins and cutaneous nerves can be seen between skin and deep fascia, branches which all penetrate the fascia. The venous network in hand and forearm is gathered up in two main channels: the *cephalic*

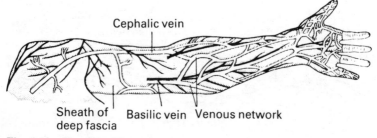

Cephalic vein

Sheath of   Basilic vein   Venous network
deep fascia

Fig. 6.19   The flayed arm, showing the subcutaneous structures, anterior aspect

vein, running up the lateral side of the biceps to the shoulder, and the *basilic* vein on the medial side, piercing the fascia halfway up the arm to run with the brachial artery.

There is a *bursa* between the olecranon process and the skin.

The muscle groups of the arm

Shoulder
The *deltoid* muscle on the outer side overlaps the front and back of the limb, in the same plane as the *pectoralis major*, which passes from clavicle and upper ribs to the upper humerus. The deltoid abducts the arm and the pectoralis adducts it. Adduction is also performed by the *teres major* and *latissmus dorsi* posteriorly, the latter being one of the great muscles of the back arising from the spinal column (see Fig. 9.3). Lying very deeply, and arising from the scapula, are the short rotators of the humerus, inserted into the tuberosities under the deltoid; they rotate the bone laterally or medially as demonstrated in Figs 6.22, 6.23, 6.24.

Upper arm
In cross-section, this is divided into anterior and posterior muscular compartments by the humerus, with a lateral and medial inter-muscular septum attached to each side of the bone (Fig. 6.25). The anterior (*flexor*) compartment contains muscles flexing the elbow, the biceps superficially and the brachialis near the bone. The *biceps* has two heads, a long tendon arising within the shoulder joint and a

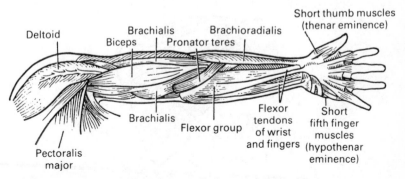

Fig. 6.20 Muscles of the left upper limb, anterior aspect

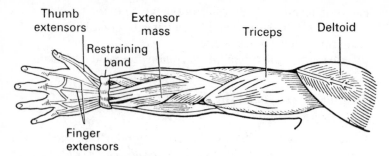

Fig. 6.21   Muscles of the left upper limb, posterior aspect

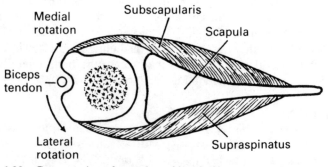

Fig. 6.22   Cross-section of scapula and head of humerus to show action of short rotator muscles

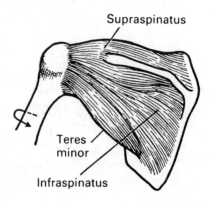

Fig. 6.23   Outward rotation by the group of short rotator muscles

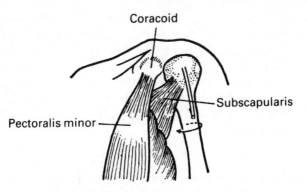

Fig. 6.24 Inward rotation by the subscapularis muscle

short one from the tip of the coracoid; they join to form the main belly, from which the tendon of insertion issues just above the elbow to reach the bicipital tuberosity of the radius. In addition to its flexing action, the biceps is the most powerful supinator muscle since it is inserted near the back of the radius and rotates it on its

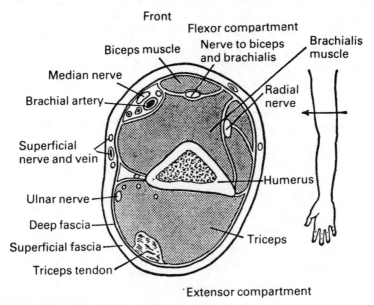

Fig. 6.25 Cross-section of the upper arm near the elbow

long axis; it is essential in powerful supination as when using a corkscrew or screwdriver. The underlying *brachialis* arises from the front of the humerus; it is an elbow flexor, inserted into the coronoid process of the ulna.

At the back of the arm, in the *extensor* compartment, the *triceps* arises by three heads, two from the humerus and one from the scapula below the glenoid. The tendon of insertion goes into the olecranon process of the ulna.

Forearm

This is divided into flexor and extensor compartments by the interosseous membrane between radius and ulna; each compartment contains a complex group of muscles with an associated nerve trunk.

In the *flexor* compartment are a superficial and a deep group of muscles, with the median nerve running between them. The *superficial* group arises from a common flexor origin on the medial humeral epicondyle. It includes the pronator teres, the main pronator of the forearm, inserted into the middle of the radius; the superficial flexors of the fingers; and the flexors of the wrist joint. The *deep* group arises from the forearm bones and interosseous membrane, and includes the deep flexors of the fingers and the long flexor of the thumb. The lateral boundary of the front of the forearm

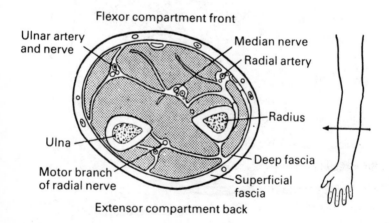

Fig. 6.26   Cross-section of right forearm, viewed from above

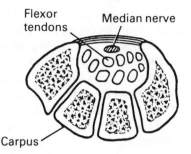

Fig. 6.27 Cross-section of wrist to show the flexor tendons in the carpal tunnel

is the brachioradialis, which passes from the lateral humeral epicondyle to the lower radius, acting alternately as pronator and supinator. All these flexor bellies become tendinous 2·5 or 5 cm above the wrist, and those tendons destined for the fingers pass into the hand with the median nerve under a transverse ligament spanning the arch of the carpal bones, an arrangement known as the carpal tunnel (see Fig. 6.27). Each finger, but not the thumb, has a superficial and a deep flexor tendon which will be studied later.

The *extensor* forearm compartment also has superficial and deep layers; and, just as the superficial flexors arise from the front of the medial epicondyle, so the superficial extensors arise from the back of the lateral epicondyle. The main nerve at the back is the radial. The groups of extensor muscles are:

1 The extensors of the wrist joint;
2 The extensors of the fingers (only one tendon to each finger, except the index and fifth, which have two);
3 The extensors of the joints of the thumb.

All these tendons are bound down by a band of deep fascia at the wrist to prevent them from bowstringing backwards.

The blood vessels of the arm

The main artery to the upper limb begins in the neck as the *subclavian*. On the right this is a branch from the innominate artery. On the left, it arises directly from the arch of the aorta in the chest.

As the vessel passes out through the axilla to the arm, it is known as the *axillary* artery and in the upper arm as the *brachial*. Here it lies medial to the biceps, close to the median and ulnar nerves. At the elbow the brachial divides into radial and ulnar branches, which course down the corresponding sides of the forearm; at the wrist the ulnar artery continues into the palm, while the radial turns on to the back of the carpus. In the palm there are two *arterial arches* formed between branches of the two vessels; from which the digital arteries to the fingers are given off. A less important *dorsal arterial arch* is formed on the back of the carpus by the radial vessel alone.

The *veins* are less clear-cut in arrangement, forming a diffuse subcutaneous network in hand and forearm, but the main lateral (*cephalic*) and medial (*basilic*) channels emerge at the elbow. The cephalic runs to the shoulder and dives under the clavicle to join the subclavian vein; the basilic vein runs with the brachial artery to the axilla, continuing as the axillary vein and ultimately as the subclavian in the neck (Fig. 6.19).

The fine *lymphatic* vessels are arranged in a superficial network corresponding to the venous pattern, and the main lymphatic trunks run with the brachial artery, entering the main group of lymph glands of the arm in the axilla. There is a small gland-station halfway up the arm, in front of the medial epicondyle, the epitrochlear gland (Fig. 5.13).

The nerves of the arm (Figs 6.17, 6.18)

The nerves of the limb are branches of a complex *brachial plexus* of spinal nerve roots situated in the lower neck, behind the clavicle, and in the axilla. From this plexus emerge the main trunks of the arm – median, radial and ulnar – which are grouped closely around the axillary vessels in the axilla.

The *radial* nerve runs round to the back of the upper arm, passing spirally round the humerus in its groove to emerge anteriorly just above the elbow in the fold between the biceps and brachioradialis muscles. It then winds back round the neck of the radius to become the main nerve of the extensor compartment of the forearm. It is mainly a motor nerve, supplying the triceps, brachioradialis, and extensors of wrist, thumb and fingers; but it has a small sensory supply to the skin on the back of the hand between thumb and index, and to the outer side of the forearm, and this branch runs with the radial artery in the flexor compartment.

The *median* nerve passes down the arm with the brachial artery and lies in the forearm between the superficial and deep planes of flexor muscles in the anterior compartment. It supplies most of the forearm flexors and enters the palm under the transverse carpal ligament with the flexor tendons of the fingers. In the hand it gives an important branch to the small thumb muscles, and sensory branches accompany the digital arteries, supplying the skin of the thumb, index, middle and lateral half of the ring fingers on their anterior aspects.

The *ulnar* nerve also runs with the brachial artery in the arm, but enters the forearm by passing behind the medial epicondyle in the ulnar groove. It runs in the flexor compartment with the ulnar artery, enters the hand superficial to the transverse carpal ligament and supplies the intrinsic muscles responsible for the fine coordination of the fingers. It also supplies sensation to the fifth and inner half of the ring fingers, on front and back.

The hand

The main *palmar space* of the hand is bounded by the thenar and hypothenar eminences on each side; its floor is the skin of the palm

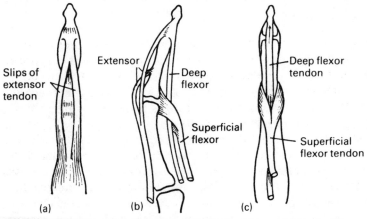

Fig. 6.28   Tendon arrangements in an individual finger: (a) Back of finger; (b) side view; (c) front of finger

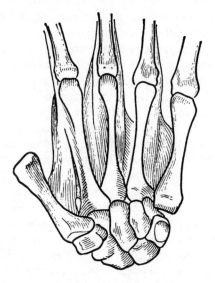

Fig. 6.29   The interosseous muscles of the hand

and its roof the metacarpals. Through it pass the flexor tendons to the fingers and in it lie the arches of the ulnar artery with their branches to the fingers, and the accompanying digital branches of the median and ulnar nerves. Between the metacarpals and arising from them are the *interossei*, the small intrinsic muscles of the hand. Their tendons wind round the metacarpal necks to be inserted into the extensor tendon expansion on the back of the proximal phalanges, and they abduct and adduct the fingers, as well as performing certain characteristic movements described in p. 110.

The *dorsum* of the hand is simply a shallow space between the skin and the backs of the metacarpals traversed by the extensor tendons of the fingers.

In the *fingers* the arrangement of the flexor tendons is complex. Each digit, save the thumb, has two flexors, superficial and deep, which are inserted into the phalanges as shown in Fig. 6.28, the deep tendon on its way to the terminal phalanx splitting the other in two. The tendons are facilitated in gliding by being enclosed in a smooth, *synovial sheath*, and each sheath is enclosed in a fibrous tunnel attached to the phalanges on each side (see Fig. 6.30). Each sheath

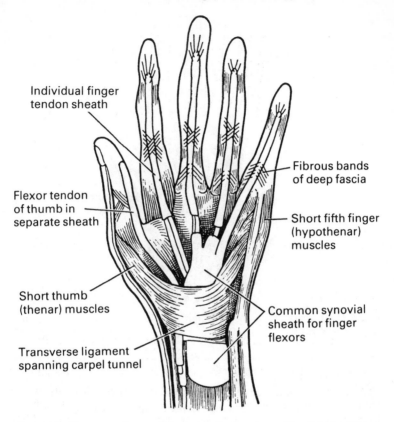

Individual finger tendon sheath

Flexor tendon of thumb in separate sheath

Short thumb (thenar) muscles

Transverse ligament spanning carpel tunnel

Fibrous bands of deep fascia

Short fifth finger (hypothenar) muscles

Common synovial sheath for finger flexors

Fig. 6.30   Flexor tendons and their sheaths in wrist and hand (After *Gray*)

is separate except that of the little finger, which expands in palm into a general sheath for all the tendons passing through the carpal tunnel.

On the dorsum, the arrangement of extensor tendons is simpler; there is a common synovial sheath as they pass over the wrist, but the individual tendons of the fingers are without sheaths. Fig. 6.28 indicates how the tendon of each finger splits up as it is inserted; note the broad extensor expansion which strengthens the capsule of the metacarpophalangeal joint, it is also the insertion of the intrinsic muscles. The main features of the *nails* are shown in Fig. 5.7.

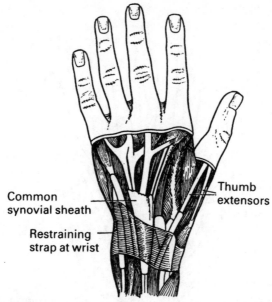

Fig. 6.31 Extensor tendons and their sheaths in wrist and hand (After *Gray*)

Function of the hand

In animals the digits are used purely for crude gripping and the long flexors and extensors alternately contract and relax their hold (Fig. 6.32(a), (b)).

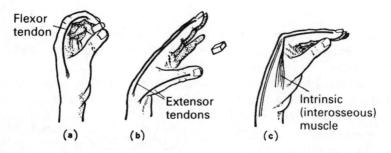

Fig. 6.32 Movements of the fingers: (a) and (b) crude movements; (c) fine use of the fingers

Man has also developed a fine coordinated action of the fingers themselves, by the evolution of local intrinsic muscles confined to the hand. Their essential action is adoption of the writing position in the fingers and opposition of the thumb to them in a fine grip. The small muscles achieve this by flexing the metacarpophalangeal joints and then extending the interphalangeal joints by pulling on the long extensor tendons (Fig. 6.32(c)).

7

Regional Anatomy: the Leg

Bones of the lower limb

Innominate bone

The arrangement of the pelvic girdle was shown in Fig. 3.15. Each half of the pelvic ring is made up of one innominate or hip bone. The two bones articulate in the midline anteriorly at the pubic symphysis, the bony prominence felt just above the external genitalia. Behind, each bone articulates with the sacrum of the spine at the sacroiliac joint, a strong junction allowing little motion.

The innominate bone itself is a support for the limbs, an attachment for the limb muscles, and a bony container for the pelvic organs (see Chapter 8). It consists of three main portions – ilium, ischium and pubis – all centred on the acetabulum, the socket for the femoral head.

The *ilium* is a flaring sheet of bone whose upper border is the iliac crest, felt with the hand on the hip. This crest has a spine at either end.

The *pubis* has a superior and inferior strut or ramus, the two rami enclosing a large gap, the obturator foramen; on the superior ramus is a little knob, the pubic tubercle, which can be felt at the inner end of the groin.

The *ischium* is the dependent portion of the pelvis carrying the broad ischial tuberosity which takes weight in sitting.

The *acetabular socket* of the hip joint is a deep cavity in the centre of the external aspect of the bone, with an encircling, rim shaped

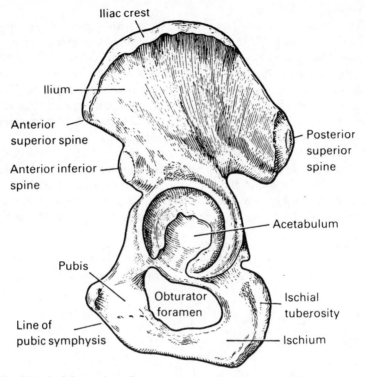

Fig. 7.1 Left innominate bone, outer aspect

like a horse-shoe, deficient at the acetabular notch in front and below.

There are important ligaments and muscles attached to the innominate bone. The anterior spine of the ilium and the pubic tubercle are spanned by the *inguinal ligament* of the groin, separating thigh from abdomen, under which the great vessels and nerves pass into the limb. The obturator foramen is bridged by the *obturator membrane*, pierced by nerve twigs and small vessels, and the acetabular notch is crossed by the *transverse ligament*. The broad outer surface of the ilium is the origin of the gluteal muscles of the buttock; the ischium, of the hamstrings of the back of the thigh; and the pubis, of the adductor muscles which bring the leg towards the midline.

Femur

This bone has a long, cylindrical shaft with slight outward and forward bowing. At its upper end the *head*, some two-thirds of a complete sphere, is set well off the shaft at an angle of 130° by the long stout *neck*. At the base of the neck are the two *trochanters*, the great trochanter laterally and the small trochanter medially. The neck acts as a lever for the muscles attached at its base, so that the restriction of motion which is the price of security is compensated by an increase in the mechanical advantage.

The shaft is thickly clothed by muscles, many attached to the bony ridge running down the length of its posterior aspect – the *linea aspera*. Not far below the neck, the posterior surface is marked by the gluteal tuberosity, the insertion of the great gluteus maximus muscle of the buttock. A few inches above the knee, the shaft

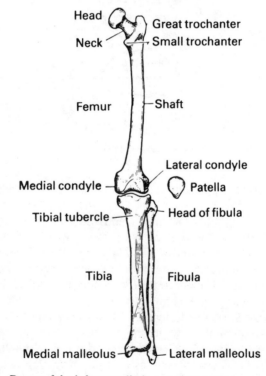

Fig. 7.2 Bones of the left upper limb, anterior aspect

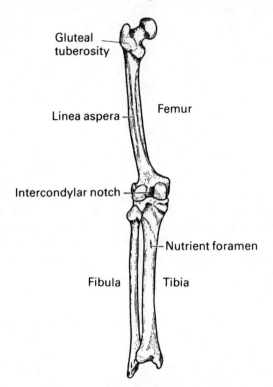

Fig. 7.3   Bones of the left lower limb, posterior aspect

expands in triangular fashion, ending in two massive *condyles*, lateral and medial, separated by a deep intercondylar notch; the medial is more prominent and is at a slight angle to the line of the shaft, as seen from below (Fig. 7.3). The lower end of the femur has a smooth surface above the condyles in front for articulation with the patella, and a broad posterior surface facing into the *popliteal fossa*, the hollow at the back of the knee. There is a small epicondyle on the outer surface of each condyle, and the adductor tubercle at the summit of the medial condyle is the lowest attachment of the adductor muscles.

The *patella* or knee-cap is a sesamoid bone, i.e. situated within tendon, in this case the tendon of the great quadriceps muscle in front of the thigh which extends the knee. It is roughly triangular –

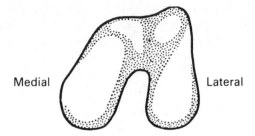

Fig. 7.4 Condyles of left femur, end-on view

apex downwards – and has an upper and lower pole, and anterior and posterior surfaces. The posterior surface articulates with the smooth surface of the femur above the condyles. The quadriceps tendon is inserted into the upper pole, and its pull is transmitted to the patellar ligament, which connects the lower pole to the tubercle of the tibia.

Tibia and fibula
These are the paired bones of the leg, on the medial and lateral sides respectively. They articulate together at the superior (proximal) and inferior (distal) tibiofibular joints. The tibia articulates with the femur at the knee and with the talus bone of the tarsus at the ankle. The fibula participates in the ankle, but is excluded from the knee joint. Each bone has an opposing interosseous border, sharp margins connected by the interosseus membrane.

The *tibia* has a long shaft, triangular in cross-section, with medial, lateral and posterior surfaces, and medial, lateral and anterior borders. The anterior border is the sharp crest that can be felt as the shin, and the flat medial surface is beneath the skin on the inside of the leg. The upper end of the bone is expanded as two broad, flat masses, the medial and lateral condyles, which articulate with those of the femur, the semilunar cartilages intervening (see p. 121). The smooth condylar surfaces are separated by a rough intercondylar area carrying a projecting tibial spine. At the upper end of the tibial crest, below the condyles, is the prominent tibial tubercle, the insertion of the extensor apparatus of the knee. The lower end of the tibia is much narrower than the upper and has a projecting process, the medial malleolus, overhanging the inside of the ankle.

The slender *fibula* transmits little body weight. Its shaft is polygonal in section, with numerous ridges for muscle attachments, and the head at the upper end articulates with the outer tibial condyle. The head carries at its apex a pointed styloid process, the insertion of the biceps muscle of the thigh and the attachment of the lateral ligament of the knee. The lower end of the bone forms the lateral malleolus, which overhangs the outside.

Foot

The bones of the foot fall into three groups: the *tarsals*, the *metatarsals* and the *phalanges* of the toes. The *talus* lies immediately beneath the long bones, articulating with them at the ankle, and rests on the *calcaneus* or heel bone. But it so inclines to the inner side of the foot that the anterior ends of talus and calcaneus lie side by side, forming the proximal row of the tarsus. The distal row

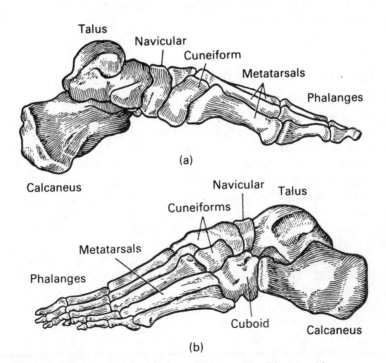

Fig. 7.5 The bones of the left foot, (a) inner side, (b) outer side

consists of three *cuneiforms* on the inner side and the *cuboid* laterally; between the two rows on the inner side of the foot is the *navicular*, separating talus from cuneiforms.

The *talus* has a *body*, with a rounded superior surface at the ankle articulation, and a *neck* carrying the *head*, which fits into the navicular. The *calcaneus* has a projecting posterior *tuberosity*, the prominence of the heel, set at an angle to the body of the bone. It articulates with the back of the cuboid. Talus and calcaneus are connected by a strong interosseous ligament.

The *cuboid* and *cuneiforms* are irregular bony masses articulating with the metatarsals. The latter are similar to the metacarpals of the hand except that the first metatarsal lies closely parallel to the others, because of the relative lack of mobility of the great toe. It is stouter than the rest and its head is supported underneath by a pair of tiny sesamoid bones. The long axis of the foot, as regards adduction and abduction of the toes, is the second metatarsal, not the middle, as in the hand.

Joints of the lower limb

Hip joint

The deeply embedded head of the femur forms the hip joint by articulation with the acetabulum. This socket is deepened by a fibrocartilaginous rim or *labrum* round its periphery, and its deep central portion is out of contact with the head and filled by a fatty pad. In the centre of the femoral head is a small depression, the *fovea centralis*, opposed to this non-articulating region of the acetabulum. From the fovea a round cord, the *ligamentum teres*, runs to the margins of the acetabular notch; it carries blood vessels to the head of the femur.

The capsule of the hip is strong and thick, a blending of bands from each portion of the innominate bone – iliofemoral, pubofemoral and ischiofemoral ligaments. It is attached proximally around the acetabular margin and reaches distally to the base of the femoral neck in the trochanteric region, i.e. the whole of the neck is inside the capsule. The synovial membrane lining the capsule is reflected upwards at the base of the neck, clothing the latter as far as the head.

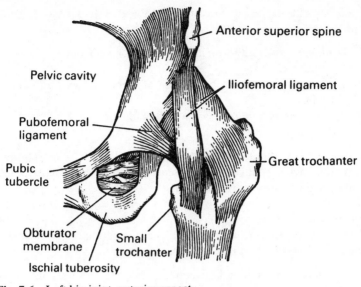

Fig. 7.6 Left hip joint, anterior aspect

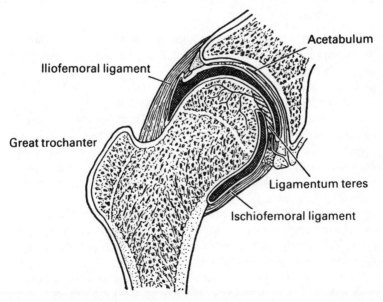

Fig. 7.7 Hip joint, longitudinal section (After *Gray*)

Movements at the hip correspond to those of the shoulder. They are:

1   *Abduction* and *adduction* in the coronal plane, from and towards the midline;
2   *Flexion* and *extension* in the sagittal plane, before and behind the trunk;
3   *Lateral* and *medial rotation*, a rolling movement of the thigh on its long axis.

The joint is liberally supplied with nerves, some of which also serve the knee, so pain from hip disease is often referred to the knee and thought to arise from that joint.

Knee joint

This joint is complex, with several communicating compartments; the fibula plays no part in it. The main parts are:

1   The *tibiofemoral* joint between the tibial and femoral condyle of each side, a semilunar cartilage intervening;
2   The *patellofemoral* articulation between back of patella and front of femur.

The knee is essentially a hinge joint, with flexion and extension as its principal movements, the strong collateral ligaments resisting any sideways strain. But there is an element of rotation, a screwing motion that locks the joint as it is finally straightened, so that in the fully extended position in standing it is very stable.

The loose and extensive anterior portion of the capsule is strengthened by the expansion of the quadriceps tendon muscle as it passes to the tibial tubercle. There is a blending of tendon and capsule, so that the tonus of the muscle guards the joint against strain, keeps the capsule taut and prevents fluid distending the capsule.

The collateral ligaments and posterior capsule are fairly taut; but the anterior capsule is loose, especially where it extends around and above the patella as the suprapatellar recess, a pouch which is necessarily baggy to allow flexion of the joint. When the knee is fully extended, the patella articulates with the femur above the condyles; in increasing flexion it lies against portions of both condyles.

The *synovial cavity* of the knee includes:

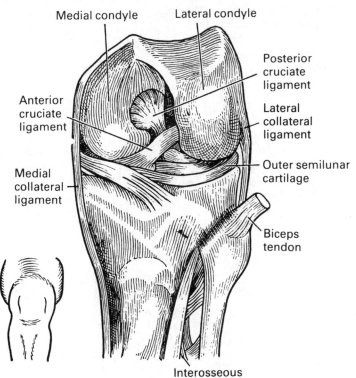

Fig. 7.8 Left knee joint, anterior aspect. The joint is bent at a right angle, and the front of the capsule has been removed (After *Gray*)

1 The *patellofemoral space*, with the suprapatellar recess in the highest part of the anterior compartment;
2 The main *anterior joint cavity* between tibia and femur on each side;
3 The *intercondylar portion*, the tunnel between the femoral condyles, traversed by the crosses of the cruciate ligaments which tie each femoral condyle to the tibia (Fig. 7.9);
4 The *postcondylar space* on each side, a loose synovial pouch behind each femoral condyle.

The *semilunar cartilages* are concentric discs intervening between the tibial and femoral condyles, and fixed to the deep aspect of the

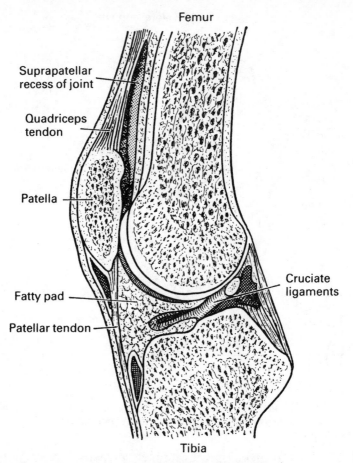

Fig. 7.9   The knee joint, longitudinal section (After *Gray*)

capsule (Fig. 7.10). They project for 1–2 cm into the joint, with a thin free edge. Their front and back ends are called the anterior and posterior horns; those of the more elliptical medial cartilage embrace those of the circular lateral disc. The two cartilages cushion the contact of the bony surfaces and are bound down at their outer margins to the upper surfaces of the tibia, with which they rotate. Nevertheless, they also have an attachment to the femur, and this double fixation is responsible for the strain which tears a cartilage.

The *tibiofibular* joints include:

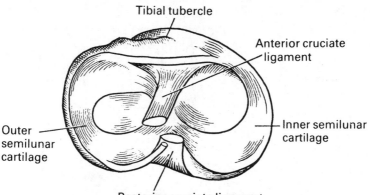

Fig. 7.10 Semilunar cartilages and cruciate ligaments of the knee joint. The view is of the upper surface of the left tibia, the femur having been removed

1 The superior tibiofibular articulation, a simple plane joint formed by the fibular head abutting against the lateral tibial condyle;
2 The connection of the shafts by means of the interosseous membrane;
3 The inferior tibiofibular joint, a firm fibrous union.

The fibula hardly moves on the tibia and is bypassed in the transmission of body weight.

Ankle joint

This is between the upper surface of the talus and the lower ends of tibia and fibula. The malleoli, overhanging the talus on each side, form a mortice in which that bone is wedged; there are strong collateral ligaments, and the only movements are those of downward plantarflexion (flexion) and upward dorsiflexion (extension).

The *tarsal joints* form a complex system best considered in terms of the movements that take place.

Movements of the foot

The foot as a whole may be *inverted* or *everted*, i.e. the sole turned inwards or outwards, and this is best understood by considering it as moving *en masse* at the subtaloid joint (the joint between talus and

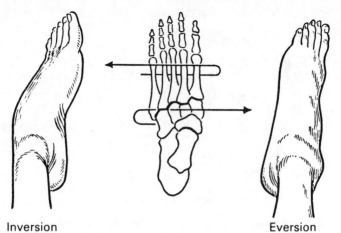

Inversion   Eversion

Fig. 7.11 Inversion and eversion of the foot. The arrows merely indicate
the direction of motion; both movements occur at the same
system of joints

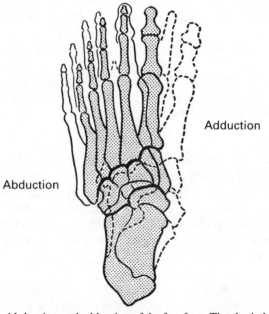

Adduction

Abduction

Fig. 7.12 Abduction and adduction of the forefoot. The shaded position is
neutral, the outlines indicate the extremes of range

calcaneus) and the talonavicular ball-and-socket, with the talus acting as a stationary pivot (Fig. 7.11).

In addition, the forefoot – metatarsals and toes – may be *adducted* or *abducted*, i.e. deviated medially or laterally, keeping the sole parallel to the ground; this occurs at the midtarsal joint, which traverses the foot and is made up of the talonavicular and calcaneocuboid articulations (Fig. 7.12).

It is not really possible to separate these movements; inversion is always accompanied by some adduction and eversion by some abduction. It is difficult to realise that none of these rotary movements is occurring at the ankle, which is capable only of up and down movement.

The remaining tarsal joints are small plane articulations; and the joints of the toes are similar to those of the fingers.

The arches of the foot
These are longitudinal and transverse. The *longitudinal arches* lie in the long axis of the foot, with a higher medial and a lower lateral long arch, with a common posterior pillar in the calcaneus. The line

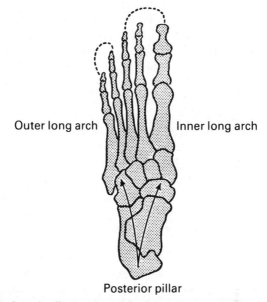

Outer long arch   Inner long arch

Posterior pillar

Fig. 7.13 Longitudinal arches of the foot

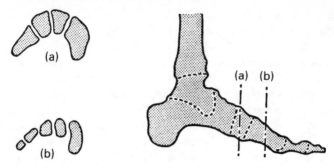

Fig. 7.14   Cross-section of the transverse arch of the foot, (a) in the tarsal, (b) in the metatarsal region

of the medial arch is calcaneus-talus-navicular-cuneiforms-inner three metatarsals. The line of the outer is calcaneus-cuboid-outer two metatarsals, this being much flatter. The talus lies at the summit of the long arches, so that the body weight constantly tends to flatten them – and would do were it not for the supporting ligaments and muscles.

The *transverse arch* is the side-to-side concavity seen in cross-section and is most marked at the bases of the metatarsals.

The long arches are not present at birth but develop in the first eighteen months; the transverse arch is present in the foetus. The importance of the arches to the function of the foot is no longer stressed. 'Flat foot' is of little importance; what matters is that the foot should be supple and capable of assuming the arched position voluntarily, as in children and ballet dancers, who have excellent function although their feet may be very flat. It is rigidity that is painful and disabling.

The supports of the arches are the binding ligaments, and the tone of muscles of the calf and sole; the latter is more important. To a certain extent the shape of the bones contributes to their maintenance. The important *ligaments* are those on the underside of the foot, short bonds between individual bones or longer structures, running from one pillar to the other.

The tendon of the *tibialis anterior* muscle runs down from the leg to pull on the medial cuneiform and thus maintain the long arch; the tendon of the *tibialis posterior* of the calf curves behind the ankle and sustains the head of the talus from below as it passes to its

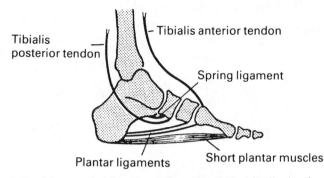

Tibialis posterior tendon

Tibialis anterior tendon

Spring ligament

Plantar ligaments

Short plantar muscles

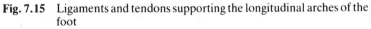

Fig. 7.15　Ligaments and tendons supporting the longitudinal arches of the foot

insertion in the navicular. And the *peroneus longus* tendon from the outer side of the leg crosses the sole of the foot from lateral to medial side, bracing the transverse arch.

Finally, the muscular mass of the sole, the short plantar muscles, attached behind to the underside of the calcaneus and extending to the toes, supports both long arches.

The ligaments are not intended to take the entire strain of body weight, except momentarily; they are guarded by those calf muscles whose tendons have been mentioned above. If this tonus is weakened, the ligaments are overstretched and acute foot strain results.

Gait and the mechanism of the foot
The arches are spring-like structures, yielding under body weight and with elastic recoil; the principal ligament, the short band between calcaneus and navicular which supports the head of the talus, is known as the *spring ligament*. In walking, the weight is planted on the heel, then transmitted along the outer border of the foot and finally across the transverse arch to the first metatarsal head; and flexion of the metatarsophalangeal joints gives a kick-off onto the other foot. The remarks on p. 110 about primitive and fine movements of the fingers also apply to the toes. In walking on rough ground, particularly with bare feet, the long flexors of the toes act as powerful gripping agents. But on pavements and in shoes it is necessary to bring the toes flat down to give an efficient thrust, and

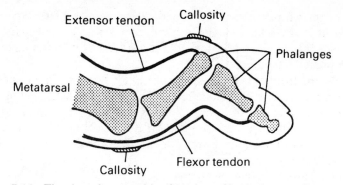

Fig. 7.16   The clawed toe resulting from loss of intrinsic control

this is accomplished by the intrinsic muscles. Should this control be lost, the unopposed long flexors and extensors produce clawing of the toes, which are hyperextended at the metatarsophalangeal and flexed at the interphalangeal joints. This exposes the metatarsal heads in the sole; and callosities form both in the sole and on the dorsum of the toes (Fig. 7.16).

Surface anatomy of the lower limb

Anterior aspect

The groin marks the transition between abdominal wall and thigh; pressure in this fold reveals the resistance of the underlying inguinal ligament, which can be traced laterally to the anterior superior spine of the ilium and medially to the small, knobby pubic tubercle; the latter is often obscured in men by the spermatic cord running to the testicle.

Immediately distal to the ligament is the *inguinal region*, which the main vessels and nerves of the limb have entered from the abdomen; the pulsation of the femoral artery can be felt just below the midpoint of the ligament, and the femoral nerve more laterally. There are numerous lymph glands.

In the *thigh*, the main anterior muscle bulge is formed by the quadriceps femoris, tapering towards its insertion into the patella. The mass on the medial side is the abductor group of muscles, whose tendons of origin can be felt running up to the pubic bone, just lateral to the external genital organs. When the knee is braced straight, a firm resistance is felt under the skin on the outer aspect of the thigh, from hip to knee; this is the iliotibial band of deep fascia, which is kept taut by the buttock muscles and helps maintain erect posture.

At the front of the *knee* is the patella, and in line with it, about 5 cm below, is the tibial tubercle; the patellar tendon connecting the two and transmitting quadriceps pull becomes obvious when the knee is extended, and a hollow on either side of the tendon indicates the anterior joint compartment immediately beneath the skin. The tibial condyles are easily seen or felt and the head of the fibula on the outer aspect; just below the head, the peroneal nerve can be rolled over the neck of the fibula.

The *tibial crest*, its anterior border or shin, is subcutaneous from tibial tubercle to ankle, with the flat subcutaneous medial surface on its inner side. On the outer side of the crest is the bulge of the muscles in the anterior compartment of the leg, and more laterally still the peroneal muscles obscure the underlying fibula, which emerges under the skin just above the ankle.

At the *ankle* the two malleoli stand out on either side, and 2·5 cm behind the medial can be felt the pulsation of the posterior tibial artery.

The *navicular* often makes a prominence on the medial border of the foot, as does the base of the fifth metatarsal on the outer side.

Posterior aspect

The great muscle masses at the back of the limb are, from above downward: the gluteal muscles of the buttock, the hamstrings in the thigh and the calf muscles of the leg.

The *buttock* has a rounded *gluteal* fold below, marking its junction with the thigh; as this is followed round laterally, the resistance of the great trochanter can be felt under the skin. Deep in the lowest point of the buttock is the apex of the ischial tuberosity.

At the back of the *knee* the tendons of the hamstrings diverge; they can be seen or felt as the biceps passes down and out to the head of the fibula, the semimembranosus and semitendinosus down and into the tibia. These tendons make the upper boundaries of the diamond-shaped space behind the knee, the *popliteal fossa*, in which lie the great vessels, which have travelled round from the front of the thigh in its lower third, as well as the sciatic nerve and its divisions. It may be possible to feel these by deep palpation.

The bulk of the *calf* is made up of the triceps, i.e. the gastrocnemius and soleus muscles, which taper to form the Achilles tendon, or heel cord, passing behind the ankle to the tuberosity of the calcaneus.

The *femoral artery* and *vein* spiral round the inside of the femur to reach the popliteal fossa. The *sciatic nerve* runs straight down the middle of the back of the thigh to divide a few centimetres above the knee, the lateral peroneal division winding round the neck of the fibula to the front of the leg.

Subcutaneous structures of the leg

Fig. 7.17 shows the flayed limb, with the superficial structures between the skin and the deep fascia. The *venous network* is gathered up into two main channels. The *great saphenous vein* runs upwards in front of the medial malleolus, along the medial border of the leg and thigh, and finally to the inguinal region, where it passes through an oval window in the deep fascia to join the main *femoral vein*. The *small saphenous vein* is formed on the lateral side of the foot, passes behind the lateral malleolus up the midline of the back of the calf as far as the knee, where it pierces the deep fascia to join the *popliteal vein*.

Both veins are accompanied by lymphatic vessels; there are a few small popliteal lymph glands at the termination of the small saphenous vein and numerous large glands surrounding the upper part of the main saphenous trunk at the groin.

The *cutaneous nerves* of the limb are many. The front of the thigh is supplied by lateral, intermediate and medial branches from the *femoral* nerve; the back by a small branch of the main *sciatic* trunk. The *saphenous* branch of the femoral nerve runs the whole length of the limb with the great saphenous vein, and the *sural* branch of the sciatic accompanies the small saphenous vein to the outer side of the foot.

The main subcutaneous *bursae* in the leg are:

1   A bursa between skin and ischial tuberosity to relieve pressure in sitting.
2   A bursa over the lateral aspect of the great trochanter.

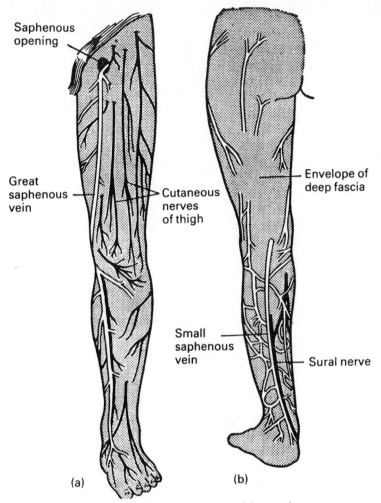

Fig. 7.17 The flayed lower limb, (a) anterior, (b) posterior

3 The prepatellar bursa in front of the patellar tendon and the lower half of the patella. This is not a part of the knee joint; it takes the weight in kneeling and its distension is the cause of 'housemaid's knee'.

4 A bursa between the skin and the Achilles tendon – when the shoe has been rubbing against the heel.

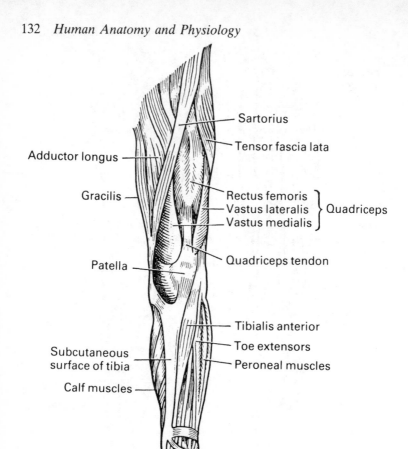

Fig. 7.18 Muscles of the left lower limb, anterior aspect

Muscles of the upper leg

These consist of the following main groups:

1 *Muscles connecting trunk and femur:* iliacus and psoas
(Fig. 7.20). These muscles flex the hip joint. The *iliacus* is a broad
sheet arising from the inner (pelvic) surface of the iliac portion of

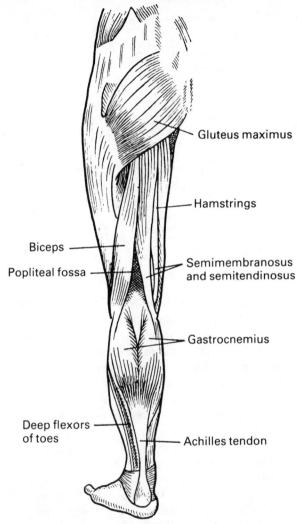

Gluteus maximus

Hamstrings

Biceps

Popliteal fossa

Semimembranosus
and semitendinosus

Gastrocnemius

Deep flexors
of toes

Achilles tendon

Fig. 7.19   Muscles of the left lower limb, posterior aspect

the innominate bone, while the *psoas* is a long belly at the back of
the abdominal cavity alongside the lumbar vertebrae. The two
muscles form a conjoined *iliopsoas tendon*, which enters the thigh
by passing under the inguinal ligament lateral to the femoral vessels
to its insertion at the lesser trochanter.

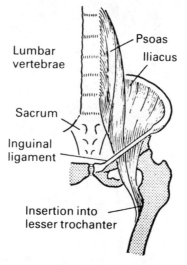

Fig. 7.20   Iliopsoas group of muscles, left side

2   *Muscles connecting pelvis and femur.* There are two main groups: the gluteals and the adductors.

The *gluteal* muscles of the buttock arise from the outer surface of the ilium and the back of the sacrum. The most superficial, covering the others, is the great *gluteus maximus*, the rump, which is inserted into the gluteal tuberosity of the femur and the iliotibial band of deep fascia. It is an important postural muscle, for it extends the hip, carries the leg back in walking and braces the limb by tightening the deep fascia (Fig. 7.19).

Beneath it are the smaller *gluteus medius* and *minimus*, which are the main abductors of the hip from the midline. These are also of postural importance for they make it possible to stand on one leg by pulling the pelvis over in line with the weight-bearing limb, and resist the tendency of the trunk to fall to the other side. They thus make walking possible, for this is no more than an alternate standing on either leg (Fig. 7.22).

The *adductors* – longus, brevis and magnus – lie on the inner side of the thigh. They arise by tendons from the pubis and ischium, running down and out to be inserted into the shaft of the femur as far down as the adductor tubercle. They adduct the femur to the midline. The three muscles are arranged in layers, from before

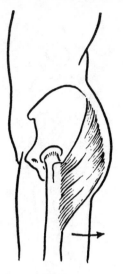

Fig. 7.21 Action of gluteus maximus

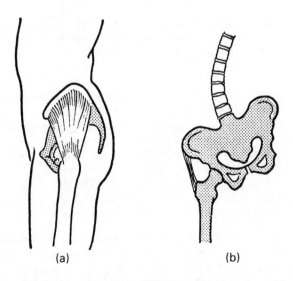

(a) (b)

Fig. 7.22 (a) The gluteus medius, (b) The stabilising action of the gluteus medius in standing on one leg

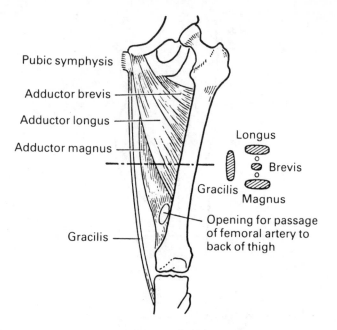

Pubic symphysis

Adductor brevis

Adductor longus

Adductor magnus

Longus

Brevis

Gracilis

Magnus

Gracilis

Opening for passage
of femoral artery to
back of thigh

Fig. 7.23 Adductor muscles of the left thigh. Their disposition is also
shown in cross-section, with the two branches of the obturator
nerve sandwiched between the layers

backwards, as named, and are supplied by the *obturator nerve*,
which enters the thigh from the pelvis after piercing the obturator
membrane. One strap-like member of this group, the *gracilis*,
reaches as far as the tibia.

3　*The short rotators of the hip* are situated deep in the buttock,
close to the sciatic nerve, and are seen only after removing the
overlying gluteal muscles. They arise from the sacrum, insert into
the base of the femoral neck and laterally rotate the hip.

4　The *quadriceps femoris* is the great mass on the front of the thigh
responsible for extension of the knee. Three heads arise from the
femur – the *vastus lateralis, intermedius* and *medialis* – and the
fourth is the *rectus femoris* from the ilium just above the acetabulum. There is a common quadriceps tendon emerging in the lower

thigh and inserted into the patella; its expansions strengthen the capsule of the knee joint.

5   The *sartorius* is a long strap-like muscle arising from the anterior superior spine of the ilium and traversing the front of the thigh superficially downwards and inwards to its insertion in the upper tibia. It helps to flex both hip and knee, and to sit with crossed legs.

6   The *hamstrings* are the bulk of the back of the thigh, a powerful group arising from the ischial tuberosity which flex the knee by inserting into the tibia and fibula. They are the *semimembranosus, semitendinosus* and *biceps* which diverge as they pass down the thigh; the first two pass inwards to the upper tibia, while the biceps, which acquires an additional head from the back of the femur, is inserted on the outside of the knee into the head of the fibula. This ∧-shaped arrangement of the tendons produces the upper boundary of the ◇-shaped popliteal fossa at the back of the knee, the lower sides of the diamond being the two heads of the gastrocnemius muscle. The sciatic nerve, which lies in the thigh under cover of the hamstrings, emerges between the upper fork to lie comparatively superficially at the apex of the popliteal region. The space is roofed by deep fascia and has the popliteal surface of the femur as its floor.

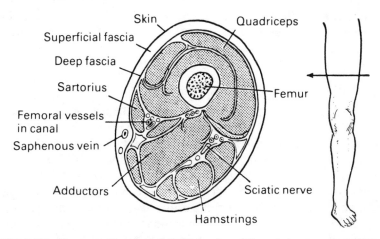

Fig. 7.24   Cross-section of the left thigh, viewed from below, to show the muscle masses

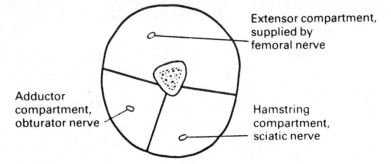

Fig. 7.25   Key to Fig. 7.24. The three compartments of the thigh, with their nerves

The disposition of these muscle groups is shown in cross-section (Figs 7.24, 7.25). The inter-muscular septa of deep fascia mark off three compartments, each containing a muscle group with its supplying nerve, which also innervates the overlying skin. The anterior or extensor compartment contains the *quadriceps*, supplied by the femoral nerve; the medial contains the *adductors*, supplied by the obturator nerve; and the posterior the *hamstrings*, supplied by the sciatic.

The sartorius in its diagonal course is applied to the side of the vastus medialis – to produce a tunnel-like space called *Hunter's canal*, through which pass the femoral vessels on their journey round the femur to emerge in the popliteal fossa, where they become the popliteal artery and vein.

Muscles of the lower leg

Cross-section of the leg reveals the main compartments and muscle groups as follows. The interosseous membrane between the bones separates a main *anterior compartment*, containing the muscles that extend the ankle and toes, from the *posterior compartment* for the calf muscles; and there is a small separate *lateral compartment* on the outer side of the fibula for the peroneal muscles which evert the foot. The main posterior compartment is subdivided into a *superficial* portion for the gastrocnemius and soleus (together forming the triceps), and a *deep* portion for the long flexors of the toes and the tibialis posterior, which inverts the foot.

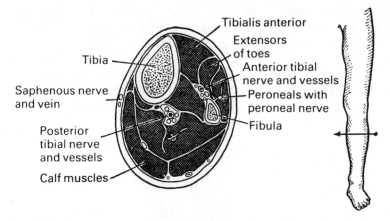

Fig. 7.26 Cross-section of left lower leg, seen from below, showing the main muscle masses

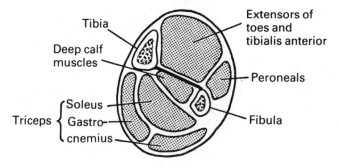

Fig. 7.27 Key to Fig. 7.26. Compartments of the lower leg

The anterior group: tibialis anterior, extensor hallucis longus, extensor digitorum longus (Fig. 7.18)

These are extensor muscles, supplied by the anterior tibial division of the peroneal nerve. Their tendons run in front of the ankle, where they are bound down by the retinacular bands of deep fascia. The tibialis anterior on the medial side, is inserted into the first metatarsocuneiform junction, where it helps maintain the inner long arch (Fig. 7.15) and dorsiflexes the foot. The extensors of the toes traverse the dorsum of the foot to be inserted like the corresponding tendons in the hand, except that in the foot there is a duplicate set of extensor tendons arising from a short extensor muscle situated on the dorsum of the foot itself.

The *peroneal* muscles, longus and brevis, evert the foot and dorsiflex the ankle; they are supplied by the peroneal nerve. P. brevis has a short course to the base of the fifth metatarsal, whereas p. longus dives into the sole and runs transversely to reach the base

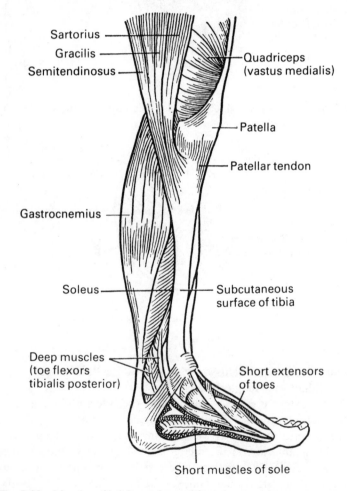

Fig. 7.28   Muscles of left lower leg and foot, inner aspect

of the first metatarsal, so helping maintain the transverse arch. Both tendons pass round the back of the lateral malleolus to reach the foot.

The posterior group (calf)

1 *Superficial layer: triceps.* The triceps is the great calf muscle acting to flex (plantarflex) the foot through its insertion, via the Achilles tendon, into the tuberosity of the calcaneus. It comprises the superficial *gastrocnemius*, with a head arising from the back of each femoral condyle and the underlying *soleus* arising from the back of the upper tibia and fibula. These form the Achilles tendon halfway down the calf.

2 *Deep layer: tibialis posterior, long flexors of the toes.* These arise from the back of the tibia and fibula and interosseous membrane. Their tendons enter the sole by passing behind the ankle and the medial malleolus. The long flexors are attached to the toes, like the flexor digitorum profundus in the arm. But whereas the fingers have two sets of flexor tendons, both arising in the forearm, those of the calf correspond only to the profundus, and the duplicate tendons are supplied by a short muscle in the sole. The tibialis posterior is inserted into the navicular and sustains the inner long arch, taking some weight off the spring ligament by supporting the head of the talus; it is a powerful invertor of the foot as well as a plantarflexor of the ankle.

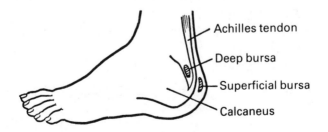

Fig. 7.29 Achilles tendon and related bursae

The foot

The general arrangement of the foot and toes, flexor and extensor tendons, their synovial sheaths and so on, resembles that of the hand, with local differences, due to the duplication of the long flexor and extensor tendons by short extensors and flexors situated in the foot itself, arising from the upper- and undersurfaces of the calcaneus.

The deep plantar fascia of the sole is very dense and covers a complex layered pattern of small plantar muscles.

The blood vessels of the leg

The *femoral artery* and *vein* are the downward continuation of the external iliac vessels from the pelvis. They pass beneath the inguinal ligament, carrying a sheath prolonged from the deep fascia of the abdomen, and lie in the groin with the vein medial to the artery. Their course in the thigh lies in three parts: in the upper third, in the inguinal region; in the middle third, in Hunter's canal, under the sartorius muscle; in the lower third they pass round the back of the femur to enter the popliteal fossa as the *popliteal* vessels. In the thigh the artery has branches to supply the muscle masses, and it divides in the popliteal fossa into an anterior and posterior tibial branch. The *posterior tibial* continues the main line of the vessel into the calf, where it lies close to the interosseus membrane, enters the sole behind the medial malleolus with the flexor tendons and supplies digital arteries to the toes. The *anterior tibial* passes forward from the popliteal fossa through the interosseous membrane into the anterior compartment of the lower leg and runs down the front of the membrane, emerging on the dorsum of the foot to give branches to the toes.

The main branches of the venous trunk are similar.

The nerves of the leg

The *femoral* and *sciatic* nerves are the great anterior and posterior trunks and, with the *obturator* and certain other twigs, are derived from the *lumbosacral plexus* of spinal nerve roots situated in the abdomen and pelvis.

The *femoral nerve* enters the thigh beneath the inguinal ligament lateral to the femoral vessels, and breaks up into the following branches:

1   The lateral, intermediate and medial cutaneous nerves of the thigh;
2   Muscular branches to the quadriceps femoris;
3   Twigs to the hip and knee joints.

The *obturator nerve* enters the adductor compartment of the thigh from the pelvis by piercing the obturator membrane; it supplies the adductors, the overlying skin and the knee joint.

The *sciatic nerve* is formed in the pelvis, on the deep aspect of the sacrum; emerges into the buttock, where it lies deeply under the glutei; and runs vertically down to the popliteal fossa, where it divides into medial (*tibial*) and lateral (*peroneal*) branches. In the upper part of the thigh it lies between ischial tuberosity and great trochanter, it is covered by the hamstrings in the midthigh and more superficial when the hamstring tendons diverge to their insertions.

The *tibial* nerve continues the course of the sciatic from the upper angle of the popliteal fossa to the lower, lying superficially under the deep fascia and crossing the popliteal vessels; it supplies the calf muscles, forming a common bundle with the posterior tibial artery, with which it enters the foot to supply the intrinsic muscles and sensory branches to the toes.

The *peroneal* branch follows the biceps tendon to the head of the fibula, winds superficially round the neck of that bone, gives a superficial branch to the peroneal muscles and the skin over the lateral side of the calf, and continues as the anterior tibial nerve with the anterior tibial artery in the extensor compartment. Here it supplies the extensor muscles, enters the dorsum of the foot and gives digital branches to the toes.

8

Regional Anatomy: the Abdomen

Abdominal cavity and boundaries

The abdominal cavity is the largest body space and more extensive than obvious at first, for its roof, the diaphragm, reaches high into the chest under the lower ribs.

The right and left domes of the diaphragm separate the right and left lungs above from the corresponding lobes of the liver below, with the heart and pericardium sitting on the flat middle part of the partition. In full expiration the right dome of the diaphragm is at the level of the fifth rib, and the left is 2·5 cm lower.

The *abdomen proper* – contains the main internal organs, bowel, liver, pancreas, spleen, the kidneys, adrenals and great vessels – and the *pelvic cavity* below, enclosed by the innominate bones and contains the terminal bowel, the bladder and genital organs. The two cavities are continuous at the pelvic inlet. A longitudinal section through the trunk shows that the main axes of the two cavities are at an angle with a backward inclination of the pelvis, whose organs are relatively separate from those in the abdomen.

The abdomen is relatively unprotected by bony framework, though this is compensated to some extent by a strong muscular belly-wall.

The boundaries of the abdomen are:

1   *Behind*, the lumbar vertebrae of the spinal column, clothed by the psoas and quadratus lumborum muscles;
2   *In front* and *at the sides*, the muscles of the flank and anterior abdominal wall;

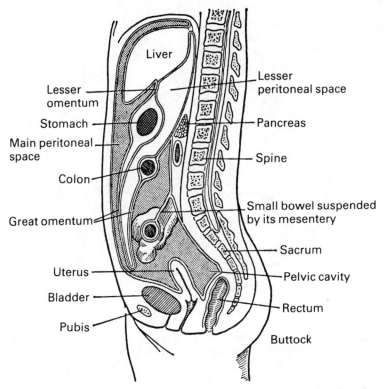

Fig. 8.1 Longitudinal section of the (female) abdomen, showing the distinction between pelvic cavity and abdomen proper

3 *Above*, by the diaphragm;

4 *Below*, on each side, the iliac fossae of the innominate bone, clothed with the iliacus muscles, support part of the abdominal contents.

The cavity is lined by a serous membrane, the peritoneum (p. 152), which also clothes most of the contained organs, the viscera; these peritoneal surfaces are normally in contact and the cavity only potential, unless air is admitted by operation or injury. A cross-section shows how the vertebral column encroaches forward so that it can be felt through the anterior abdominal wall, leaving a bay on either side in which lies the kidney (see Fig. 8.2).

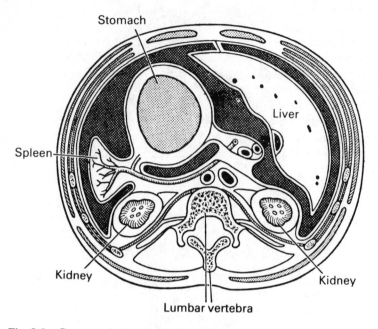

Fig. 8.2 Cross-section of the abdominal cavity. Note the recess on either side of the vertebral column in which each kidney lies. The large black space is the greater sac of the peritoneal cavity; the lesser sac is the small space immediately behind the stomach

There is considerable respiratory excursion in the abdomen; on inspiration the diaphragm is depressed as the lungs expand, and depresses the abdominal organs, the liver descending 5–8 cm with a deep breath. The cavity also varies in size with contraction and relaxation of its muscular walls and the degree of distension of the hollow viscera.

Surface anatomy

Anterior abdominal wall

On each side superiorly the *costal margins*, the lower borders of the ribs, form a ∧-shaped angle. At its apex is the tip or *xiphoid process* of the sternum. The *umbilicus*, or navel, is in the midline, on a level with the fourth lumbar vertebra. In the lowest part of the midline anteriorly is the *pubic symphysis*, formed by the junction of the two halves of the pelvis. Xiphisternum, umbilicus and symphysis are connected by a tough strip of deep fascia, the *linea alba*, a firm central attachment for the muscles, whose aponeuroses intersect at this point. When they contract, the linea alba is seen as a central depression between the *rectus* muscle on either side; the belly of the latter is crossed by two or three intersections.

Inferiorly, the symphysis can be traced out to the pubic crest on each side; the *inguinal ligament*, spanning from pubic tubercle to anterior superior iliac spine, marks the junction between abdomen and thigh; from the iliac spine the *iliac crest* can be followed round to the back.

The surface of the anterior abdominal wall is divided into zones for reference (Fig. 8.3). In the upper zone lies the epigastric region centrally, with the hypochrondriac regions on either side; in the

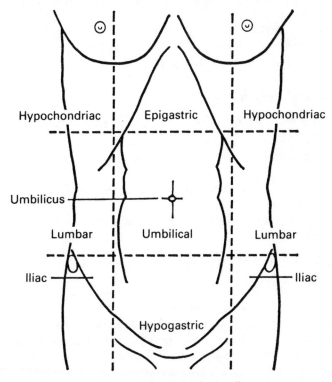

Fig. 8.3 The zones of the anterior abdominal wall

middle zone the umbilical region lies centrally, flanked by the lumbar regions; below the hypogastric region lies centrally, and has the iliac regions to right and left.

Posterior abdominal wall

In the midline of the back the spinous processes of the lumbar vertebrae are felt between the thorax and the sacrum; and on each side of the spine is the belly of the great *erector spinae* muscle. The short twelfth rib on each side marks the lowest limit of the bony thoracic cage, while below the iliac crests can be traced back towards the sacrum, ending in the posterior superior iliac spines, marked by a dimple in the overlying skin. On each side of the spine, beyond the lateral border of the rector spinae muscle, is an unprotected portion of the abdominal wall, the loin or lumbar region, between twelfth rib and iliac crest. However, the flank musculature is extremely strong.

The colon and kidneys, because they are *behind* the peritoneum, are relatively fixed – unlike the mobile liver and small bowel, which are slung freely in the peritoneal cavity.

Abdominal muscles

These are more easily drawn than described and fall into the following groups:

1   The muscles of the posterior abdominal wall: *psoas, quadratus lumborum*;
2   The muscles of the flanks: *internal* and *external obliques, transversus abdominis*;
3   The muscles of the anterior abdominal wall: *rectus abdominis, pyramidalis*.

The cross-section in Fig. 8.4 gives a general idea of the interrelation of these groups.

Muscles of posterior abdominal wall

The *psoas* has already been encountered in connection with the hip joint (p. 134); a long belly beside the lumbar vertebrae, it skirts the pelvic brim to enter the thigh. The *quadratus lumborum* lies immediately lateral to it, a quadrilateral plate between twelfth rib and iliac crest.

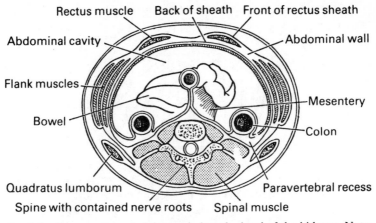

Fig. 8.4 Cross-section of abdomen below the level of the kidneys. Note the suspension of the small bowel by its mesentery from the posterior abdominal wall; also the relations of the muscle groups

Flank muscles

These are arranged in layers with the *internal oblique* sandwiched between the *external oblique* superficially and the *transversus abdominis* deeply; their fibres cross in different directions, adding to their protective power. They arise from the lower ribs above, the iliac crest below and, via a dense sheet of lumbar fascia, from the tips of the lumbar transverse processes behind. The external oblique fibres run down and in like a man putting his hand in his pocket; the internal oblique fibres are at right angles to this and those of transversus run straight round.

Near their origins, above, below and behind, these muscles are very fleshy, but as they curve round the flank to the anterior abdominal wall they merge into aponeurotic sheets extending between costal margin and iliac crests. These aponeuroses are intimately related to each other and to the rectus abdominis muscle, as indicated below; they intersect at the linea alba in the midline.

The aponeuroses, the rectus sheath, nerves and vessels of the abdominal wall

About halfway between the flank and the linea alba, the flank muscles are continued forward as aponeurotic sheets. As they

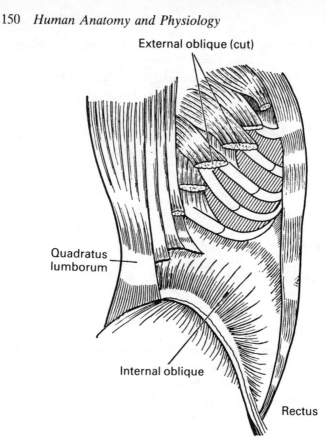

External oblique (cut)

Quadratus lumborum

Internal oblique

Rectus

Fig. 8.5 The flank abdominal muscles of the right side, intermediate layer. The external oblique has been removed, and the internal oblique is seen extending between the quadratus lumborum behind and the rectus abdominis in front

approach the lateral border of the rectus muscle, the external oblique aponeurosis passes in front of that muscle and the transversus aponeurosis behind, while that of the internal oblique splits, one layer joining each of the others (Fig. 8.4). This provides a continuous rectus sheath, whose anterior and posterior walls rejoin on the inner side of the rectus to form the linea alba.

The nerves and vessels of the anterior abdominal wall are the lower six pairs of thoracic vessels and nerves; as the lower ribs

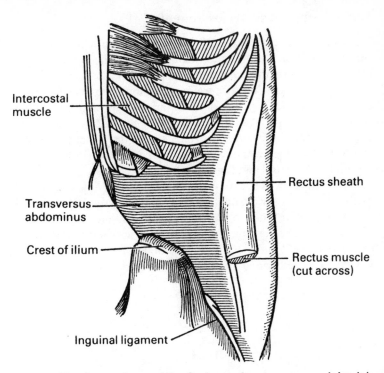

Fig. 8.6 The deepest layer of the flank muscles, transversus abdominis (After *Gray*)

overhang the abdominal cavity and are incomplete anteriorly, the neurovascular bundles in the rib spaces emerge between the flank muscles of the abdominal wall. They run between the transversus and internal oblique to the lateral border of the rectus, enter the sheath behind the muscle and finally turn forward at its inner border, piercing the linea alba to reach the skin. In their circular course, branches have been given off to the muscle layers.

Finally, within the rectus sheath, a vertical arrangement of arteries and veins is formed behind the rectus muscle by anastomosis between the superior epigastric vessels coming down from the thorax and the inferior epigastrics running up from the external iliac trunks below (Fig. 8.7).

Peritoneum, mesenteries, omenta

The abdominal cavity is a closed sac, lined with the peritoneal serous membrane; the membrane clothing the deep aspect of the abdominal wall is the *parietal* peritoneum; that reflected over the contained viscera is the *visceral* peritoneum. The parietal layer is attached to the deep surface of the anterior and posterior abdominal walls, the undersurface of the diaphragm and the upper surface of the pelvic floor. Its smooth surface allows the structures to glide freely, and a loose layer of connective tissue intervenes between the parietal peritoneum and the abdominal wall.

The main peritoneal cavity of abdomen and pelvis is known as the *greater sac*, in contradistinction to a smaller recess, the *lesser sac*, which lies behind the stomach. These spaces, and the relations of the organs to the peritoneum, are complex and best understood by studying Figs 8.1 and 8.2. Most of the organs lying in the abdominal cavity have developed in the embryo from the posterior wall,

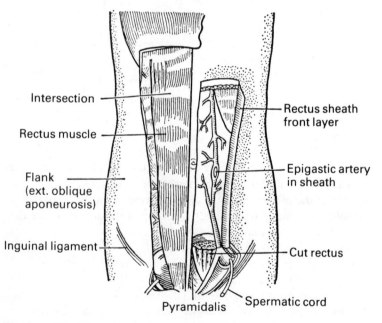

Fig. 8.7   The left rectus sheath opened and the muscle removed to show the epigastric vessels

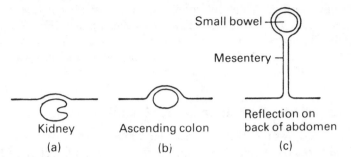

Fig. 8.8 Different degrees of peritonisation of abdominal organs. (a) A retroperitoneal organ. (b) Organ with peritoneum on three sides. (c) Freely suspended organ with mesentery

to which they still retain an attachment. The main points are (see Fig. 8.8):

1   Certain organs are entirely retroperitoneal and therefore fixed, e.g. the kidneys and pancreas; they lie on the posterior abdominal wall, with a peritoneal covering, if any, on their anterior surfaces only.

2   Other organs, the ascending and descending portions of the colon and the rectum – parts of the large bowel – have a peritoneal covering in front and at the sides; they are still retroperitoneal but with rather more mobility.

3   The small intestine, and the transverse and pelvic portions of the colon, hang freely; to do so, they have pulled out a double-layered sheet of peritoneum from the posterior abdominal wall which is shown as a *mesentery*. This is laden with fat, lymph glands and vessels, and the blood-vessels reach the bowel by running between its layers from the great vessels on the posterior abdominal wall. Small bowel, transverse and sigmoid colon each have their own mesentery; but the main one is the great sheet supporting the small bowel, known simply as *the* mesentery, whose line of attachment crosses the lumbar spine obliquely from left to right from above downwards (Fig. 8.4).

4   The stomach has two special mesenteries known as *omenta*. The *great omentum* hangs down from its lower border as an apron-like fold in front of the small bowel and then turns up to embrace the transverse colon before its final reflexion onto the posterior ab-

dominal wall, i.e. the mesentery of the transverse colon is really continuous with the great omentum of the stomach (Figs 8.1 and 5.8). The *lesser omentum* connects the upper border of the stomach to the liver, and forms the anterior boundary of the lesser sac (Fig. 8.1).

The general disposition of the abdominal organs

Fig. 5.8 shows the structures revealed when the anterior abdominal wall is removed. The *liver* occupies the upper right portion of the cavity, emerging a little below the costal margin and occupying a little of the left side, with its intermediate portion lying in the epigastric region. Beneath its lower border, on the right side, projects the *gall-bladder*. The *stomach* is largely under cover of the left ribs, but part of its anterior surface is seen in the angle between the lower border of the liver and the left costal margin. The *great omentum* descends from the lower border of the stomach, covering the *transverse colon*, which lies immediately below, and the coils of

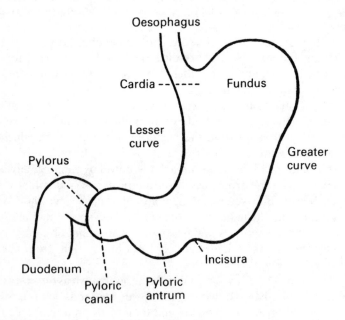

Fig. 8.9   The stomach, anterior surface

small bowel. These last are the main contents of the cavity, insinuating themselves into its recesses and dropping into the pelvis. The *spleen* is barely seen as it is tucked under the left costal margin. The more superficial viscera will now be considered.

Stomach

The stomach is the widest part of the digestive tract, connecting the oesophagus (gullet) with the duodenum, the beginning of the small intestine. The oesophagus joins the stomach shortly after entering the abdomen through the diaphragm, and the duodenum leaves the organ at the pyloric orifice. The stomach lies in the epigastric and left hypochondriac regions, but there is considerable variation with posture, digestive state and emotion. It is J-shaped, with anterior and posterior surfaces, and upper and lower borders known as the lesser and greater curvatures respectively. The lesser curve sweeps from the oesophageal entry at the cardia on the left extremity to the departure of the duodenum at the pylorus at the right-hand end. The main subdivisions are:

1   The dome-shaped *fundus*, the receptive portion, often distended with air;
2   The main *body*, concerned with digestion, marked off by a definite notch;
3   The *pyloric* or expulsive portion, narrowing as the pyloric antrum to the pyloric canal.

Both cardiac and pyloric orifices are surrounded by *sphincters*, muscular rings that are normally contracted, keeping the openings closed until relaxation is required for the entry or exit of food.

The stomach is made up of several layers (Fig. 8.10): the smooth outer *serous* or peritoneal coat; the intermediate *muscular* wall, which contains circular, longitudinal and oblique fibres; and the internal lining *mucous membrane*, a velvety, folded layer whose glands secrete the gastric juice.

The *anterior surface* of the stomach is under cover of the left lobe of the liver and the left costal margin, with a small intermediate portion on the back of the anterior abdominal wall. The *posterior surface* lies on a 'stomach bed' (Fig. 8.11), formed by organs of the posterior abdominal wall – pancreas, left kidney and suprarenal, and spleen. Between the posterior surface and its bed is the lesser

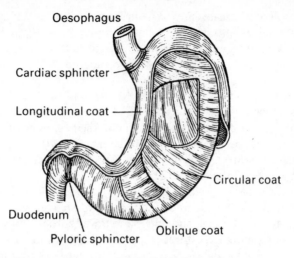

Fig. 8.10 The muscle coats and sphincters of the stomach

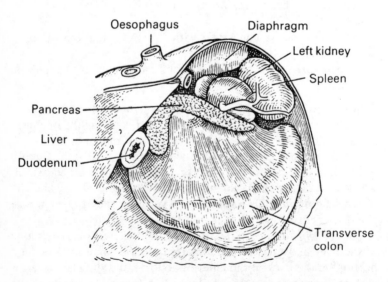

Fig. 8.11 The stomach bed. The left lobe of the liver has been cut away, and the stomach itself removed, to show the organs on which it lies (After *Gray*)

sac. The *fundus* lies in contact with the undersurface of the left dome of the diaphragm.

Small intestine

This is the portion of the digestive tract between the pyloric orifice and the large bowel. Some 6 m long, it is divided into:

1   The *duodenum*, a short coil of 25 cm immediately continuous with the stomach and bound down tightly to the posterior abdominal wall;

2   The *small intestine proper*, suspended in coils from the posterior wall by its mesentery.

The *duodenum* is C-shaped, embracing the head of the pancreas. A short first part runs horizontally from the pylorus under cover of the liver and gall bladder; the vertical second part descends in front

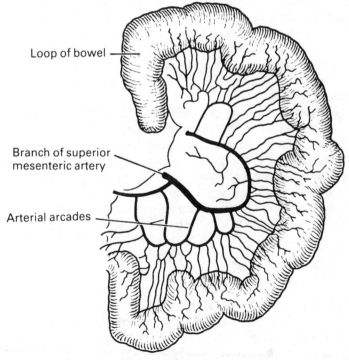

Fig. 8.12   A loop of small bowel, with its blood-vessels

of the right kidney; the third part crosses in front of the lumbar vertebrae, from which it is separated by the great vessels – aorta and inferior vena cava. It joins the small intestine at the duodenojejunal flexure. The pancreatic and bile ducts from the liver have a common opening into its second part.

The *small intestine* runs from the duodenojejunal flexure to the ileocolic valve, which marks its junction with the caecum of the large bowel. Its coils are completely surrounded by peritoneum, except for a narrow strip, the mesenteric border, where the two layers of the mesentery diverge to enclose them. The small bowel resembles the other parts of the intestine in possessing an outer *serous* coat, a *muscular* wall and a lining *mucosa*. But characteristic are the circular folds projecting into the lumen; the enormous number of minute fringes or *villi* of the mucosa, which give it a velvety appearance, a device for increasing the absorptive area; and the scattered patches of lymphoid tissue. Networks of blood and lymph vessels and nerves form plexuses between the layers, all forming larger trunks which run into the mesentery. The blood vessels of the bowel enter the root of the mesentery as the superior mesenteric branches of the aorta and vena cava, and form a pattern of arcades between its layers, from which the ultimate twigs are given off to the bowel.

The *jejunum* is the upper two-fifths of the small bowel, the *ileum* the lower three-fifths. The ileum is thinner, narrower, less vascular and contains more lymphoid tissue as digestion and secretion are preponderant at the upper end of the bowel, absorption at the lower. The small gut as a whole is enclosed by the limbs of the colon. The jejunum is above and to the left, while the ileum is central and inferior, some of its loops spilling into the pelvis before rising again as the terminal ileum to join the large bowel.

The mesentery
The leaves of the mesentery enclose:

1   Fat and connective tissue;
2   Lymph glands receiving the bowel lymphatics or *lacteal* vessels, containing milky fat-laden fluid;
3   The mesenteric vessels;
4   Nerves supplying the intestine.

The posterior attachment or 'root' of the mesentery is an oblique line of some 15 cm running down the posterior abdominal wall from the left of the second lumbar vertebra to the right sacro-iliac joint below, and crossing the duodenum and great vessels. The great disparity between this short origin and its extensive attachment to the bowel causes the membrane to be thrown into fan-shaped folds.

Large bowel (Fig. 8.13)
The large intestine runs from the end of the ileum to the external orifice of the anus. Only some 1·5–1·8 m long, it includes:

1 The *caecum*, with its *vermiform appendix*;
2 The *ascending, transverse* and *descending* limbs of the colon;
3 The *pelvic colon*;
4 The *rectum* and *anal canal*.

Only the transverse and pelvic colon have mesenteries and are freely mobile; the rest is retroperitoneal; the lower rectum and anus

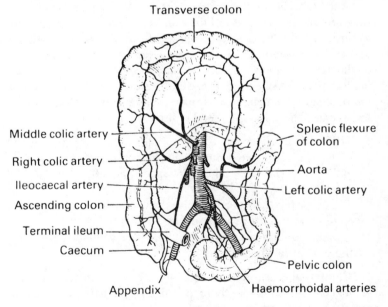

Fig. 8.13 The large bowel and its blood-vessels. The transverse loop of colon, which is normally dependent, has been turned up for clarity

are entirely below the level of the peritoneum in the depths of the pelvis.

The *caecum* lies in the right iliac fossa, the shallow basin formed by the iliac portion of the right innominate bone. It is the blind sacculated commencement of the large bowel, the terminal ileum opening into it by an aperture guarded by the ileocolic valve. The appendix is attached a little lower in the angle between. The caecum has no mesentery, but is completely invested by peritoneum, and has the mobility of a flapping balloon.

The *appendix* is a worm-like tube, blind at its free end and with a tiny mesentery of its own. Its position is far from constant; it may lie behind the caecum, hang over the pelvic brim, or turn up before or behind the terminal ileum.

Colon

The ascending, transverse and descending colon are arranged round the abdominal cavity like three sides of a square – ⬓ – but the two vertical limbs lie in the paravertebral gutter on each side and therefore behind the plane of the transverse limb, which sags down towards the pelvis. The junctions of transverse colon with the vertical limbs at either end are known as flexures, the *hepatic flexure* on the right under cover of the liver and the *splenic flexure* on the left in relation to the spleen. Because of the greater bulk of the liver on the right side, the hepatic flexure is several centimetres lower than the splenic. The colon in general is distinguished by:

1   Three superficial bands of longitudinal muscle standing out under the serous coat and traversing the bowel from end to end – the *taenia coli*, which are spaced equally round its circumference;
2   The scattered fatty tags or polyps which project as the stalked *epiploic appendages* on the surface;
3   A segmentation into coarse *sacculations* vaguely resembling those of an earthworm.

The internal structure is like that of the bowel in general, but the serous coat is incomplete where the colon is only partly peritonised, and the mucosa is smooth and pale.

The *ascending colon* lies on the posterior abdominal wall, mainly on the quadratus lumborum muscle, and is relatively immobile.

The *hepatic flexure* lies in front of the lower pole of the right kidney and is overhung by the right lobe of the liver.

The *transverse colon* arches across the abdomen, lying below the stomach and obscured by the great omentum. It is very mobile since it has a mesentery, the *transverse mesocolon*, derived, from the doubling back of the layers of the great omentum, which diverge to enclose the bowel and rejoin to form its mesentery as they pass to the posterior abdominal wall (Fig. 8.1).

The transverse colon lies in front of most of the structures of the posterior abdominal wall at this level – duodenum, pancreas, great vessels and portions of the kidneys on either side – and the prominence of the lumbar spine behind arches the colon forwards.

The *splenic flexure* lies in front of the left kidney, immediately below the spleen, with the diaphragm behind.

The *descending colon*, like the ascending, has no mesentery and is plastered onto the posterior abdominal wall behind the peritoneum. It runs down from the splenic flexure as far as the pelvic brim on the left side, where the shallow left iliac fossa gives way to the true pelvis, crossing in front of the left psoas muscle, and the left common iliac artery and vein.

The *pelvic colon*, the continuation of the descending colon, extends from the pelvic brim to the beginning of the rectum, opposite the middle of the sacrum in the depths of the true pelvis. It has a long mesentery, the pelvic mesocolon, and is loose and mobile.

The *rectum* and *anus* will be described in connection with the pelvis (p. 175).

Blood-vessels of the large bowel (Fig. 8.13)
Each limb of the colon has its main artery; the ascending and transverse colon have the *right* and *middle colic* vessels – branches of the superior mesenteric artery, which has already supplied the small bowel – while the descending limb receives the *left colic* division of another branch of the aorta, the *inferior haemorrhoidal* artery, which passes down to the rectum after giving twigs to the pelvic colon. The right, middle and left colic arteries approach the bowel, fork into two main branches which run parallel with the bowel to anastomose (connect) with adjacent vessels. They thus form a continuous arterial channel whose branches form a network

whose twigs ultimately reach the intestine. The pattern of the *veins* is similar; but, whereas the arteries have sprung from the *aorta*, the veins do not return to the companion inferior vena cava but enter an entirely separate venous trunk behind the pancreas, the *portal vein*, which gathers up the splenic vein to run up in the lesser omentum and enter the liver. This so-called 'portal circulation' ensures that the venous blood from the bowel, containing the products of digestion, passes through the liver before entering the general circulation.

Liver

This is a bulky solid organ – the largest gland in the body – in the upper right portion of the abdominal cavity. It has a brownish friable substance, with a smooth peritoneal coat, and is slung under the diaphragm by suspensory ligaments, which are the peritoneal reflexions. The large *right lobe* and the smaller *left lobe* lie beneath the right and left domes of the diaphragm, and most of the organ is under cover of the ribs. It reaches almost to nipple level on each side, and projects a little beyond the costal margin in the epigastric and right hypochondriac regions, where it touches the back of the anterior abdominal wall.

Because of the upward bulge it produces in the floor of the thoracic cavity, the lower borders of the lung and pleural cavity actually surround the upper part of the liver. The organ is ▽-shaped

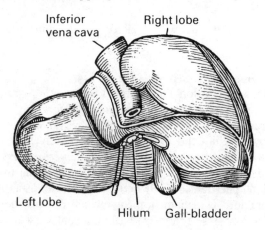

Fig. 8.14   The liver, seen from behind (After *Gray*)

in sagittal section, the sloping limb facing backwards, with superior, anterior and posterior surfaces, and bluntly rounded borders, of which the inferior is the sharpest.

The *anterior surface* is behind the lower ribs and the anterior abdominal wall in the epigastric region. Demarcation of right and left lobes is by the *falciform ligament*, which attaches the liver to the back of the anterior abdominal wall in the midline between xiphisternum and umbilicus. The *superior surface* rests on the diaphragm. The *posterior surface* is more complex. On the right it overlies the right kidney and the hepatic flexure of the colon; on the left it lies on the stomach, the upper pole of the left kidney and suprarenal gland and the spleen.

The *gall bladder* is attached to the back of the right lobe and can be seen from in front projecting below the inferior border. The *inferior vena cava* is embedded in the back of this lobe, on its way to pierce the diaphragm and enter the thorax.

At the centre of the back of the liver, between the lobes, is the root or *hilum*, where the main vessels and ducts enter and leave the organ. These are:

1   The *hepatic artery* from the aorta, dividing into right and left branches to the two lobes;
2   The *portal vein*, carrying nutriment from the bowel, dividing into two branches;
3   The right and left *hepatic ducts*, conveying bile from each lobe and joining to form the *common hepatic duct*. The *cystic duct* from the gall bladder joins the common duct a little below its formation, and the main channel thus formed is the *common bile duct*.

These structures – bile duct, hepatic artery and portal vein – run between the layers of the lesser omentum, which is attached to the hilum of the liver at one end and the lesser curve of the stomach at the other.

The *functions* of the liver include the preparation of carbohydrates and proteins for utilisation by the body after absorption from the bowel, the storage of carbohydrate as glycogen and secretion of bile. The latter is stored in the gall bladder, a reservoir with no secretory function of its own. It has a muscular coat which contracts in response to the entry of food into the duodenum; the bile empties

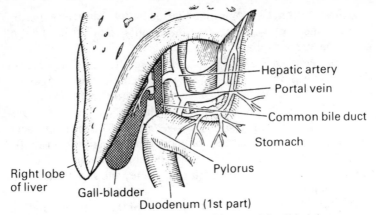

Fig. 8.15   The liver, seen from in front, with part of the right lobe cut away to show the structures entering and leaving the hilum (After *Gray*)

through the common duct into the duodenum, where it emulsifies the fat of the food into tiny globules for easier digestion.

Spleen
This lies just below the left dome of the diaphragm, corresponding to the right lobe of the liver on the opposite side; it is a much smaller organ, entirely under cover of the ribs, soft and pulpy, with a fibrous capsule. It is pyramidal or tetrahedral, with a large, smooth, convex surface under the diaphragm, and three smaller surfaces facing inwards and converging on the *hilum*, where the splenic artery and vein are attached. One surface forms part of the stomach bed (Fig. 8.11); the others are in contact with the left kidney and the splenic flexure of the colon; the tail of the pancreas reaches to the hilum.

The spleen is a reservoir of red blood corpuscles; there are muscle fibres in its capsule and it can squeeze more cells into the circulation if effort or lack of oxygen make it necessary. It is also a site for the formation of the lymphocytes of the blood and immune bodies.

Posterior abdominal wall and related structures

Looking into the abdominal cavity from the front after removal of the small bowel, we see the *posterior abdominal wall*, a bony and

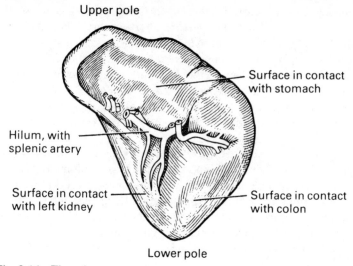

Upper pole

Surface in contact
with stomach

Hilum, with
splenic artery

Surface in contact
with left kidney

Surface in contact
with colon

Lower pole

Fig. 8.16 The spleen

muscular boundary partly obscured by certain structures. These are
the pancreas and duodenum, stretching from side to side; the
kidneys and ureters, one in each paravertebral recess; and the great
vessels running downwards in the midline.

The *posterior wall* is composed in the midline of the bodies of the
lumbar vertebrae, which project forward leaving *paravertebral
recesses* on each side in which lie the kidneys and the ascending or
descending colon (Fig. 8.2). On each side of the bodies is the psoas
muscle and, more laterally, the flat quadratus lumborum muscle,
between twelfth rib above and iliac crest below. The plexus of
lumbar nerve roots, which emerge between the vertebrae to form a
network whose branches supply the leg, lies embedded in the psoas
muscle. The upper limit of the posterior wall is the lowest attachment of the diaphragm behind; the lower limit is the pelvic brim;
laterally it is continuous with the flank muscles.

Great vessels: aorta and inferior vena cava
The *abdominal aorta* is the continuation through the aortic aperture
of the diaphragm, at the level of the twelfth thoracic vertebra, of the
thoracic portion of this great artery which has arisen from the left

ventricle of the heart. It runs down the midline in front of the lumbar bodies, as far as the fourth lumbar vertebra, where it divides into the right and left common iliac arteries, at a level just below and to the left of the umbilicus. It therefore lies very deeply, and is crossed by the pancreas and the third part of the duodenum. There is a mass of sympathetic nervous tissue on each side of its commencement, the *coeliac ganglia* or *solar plexus*, branches of which form an *aortic plexus* around the vessel. The main *sympathetic chain* lies on the vertebral bodies to either side.

The *inferior vena cava* is a great vein lying immediately to the right of the aorta, draining blood from the lower abdomen and legs. It is formed by the union of the two common iliac veins in front of the fifth lumbar vertebra, the junction lying behind the right common iliac artery. And it runs up the posterior abdominal to pierce the diaphragm and enter the right atrium of the heart.

The branches of the aorta include:

1   Single unpaired branches in the midline: the *coeliac axis* to liver, spleen and stomach; the *superior mesenteric* to small bowel, ascending and transverse colon; and the *inferior mesenteric* to the remainder of the large bowel.

2   Symmetrical paired branches on each side to the diaphragm, suprarenals, kidneys, testicles or ovaries. The largest are the *renal arteries*, and the right renal has to pass behind the vena cava to reach its kidney. Since the ovaries lie in the pelvis, and the testicles even lower, their vessels have to run down on the posterior abdominal wall, in company with the ureters. The branches of the vena cava correspond to those of the aorta, with adjustments for the difference in position – e.g. the left renal vein is much longer than the right and has to cross in front of the aorta. But the venous blood from the bowel has to pass through the liver before entering the general circulation so that the products of digestion may be dealt with. This is achieved by a secondary *portal circulation*. The superior and inferior mesenteric veins join behind the pancreas to form a main *portal vein*, which also drains the stomach and spleen. This trunk ascends in the lesser omentum to the liver, giving a branch to each lobe; in the omentum, it forms a common bundle with the bile duct and hepatic artery. After the portal blood has passed through the liver, it re-enters the general venous stream by a number of little

hepatic veins opening directly into the vena cava on the back of the liver.

The main *lymphatic* drainage from the abdomen, including the lacteal vessels carrying digested fat from the bowel, is into a small vessel, the *cisterna chyli*, lying between the upper parts of aorta and vena cava. It enters the thorax with the aorta to become the *thoracic duct*, which travels up to the root of the neck on the left side to empty into the venous circulation.

Pancreas

This is a soft, solid, lobulated organ of ⊂-shape, embraced in the concavity of the duodenum and stretching across the posterior abdominal wall and great vessels. It is composed of a *head*, the rounded right-hand extremity, grasped in the duodenal curve; a *neck*; the *body*, which stretches across to the left side, in front of the great vessels and the upper part of the left kidney; and a *tail*, turning up the hilum of the spleen.

The pancreas has important external and internal secretions. The *external secretion*, the pancreatic juice, aids digestion; it is discharged through the pancreatic duct into the second part of the duodenum. This duct forms a common channel with the common bile duct so that bile and pancreatic juice are discharged simultaneously when required; the opening is closed between meals by a sphincter.

The *internal secretion*, from island-groups of cells, is *insulin*, which passes into the bloodstream and is essential for tissue utilisation of sugar. Deficiency of this secretion is the cause of diabetes (see also p. 254).

The superior mesenteric vessels emerge from behind the body of the pancreas and then pass in front of its head to enter the mesentery of the small bowel. The *splenic artery* runs along the upper border of the organ from the coeliac branch of the aorta to the spleen. The lower parts of portal vein, hepatic artery and common bile duct are situated in or behind the head and neck.

Kidneys and ureters

The kidneys remove waste products and excess water from the blood brought to them by the renal arteries, and the urine they excrete is passed into ducts, the *ureters*, which carry it to the bladder (see Fig. 5.11). Each kidney lies in a paravertebral recess. They are bean-shaped organs, convex at their outer borders and concave toward the midline, with a body and upper and lower poles. Perched on the upper pole is the *adrenal gland*, an organ of internal secretion.

They lie obliquely, with the upper pole nearer the midline. The right kidney is 2·5 cm or more lower than the left, owing to the bulk of the right lobe of the liver. The left kidney rests on the eleventh and twelfth ribs, but the right is only on the twelfth. At the middle of the concave medial border is a depression, the root or *hilum*, where the artery, vein and ureter are attached.

Longitudinal section (Fig. 8.17) shows an outer solid substance enclosing an inner cavity, the *renal pelvis*, which collects the urine. The renal substance has an outer rind or *cortex* and a deeper *medulla*, and the latter is arranged in pyramidal masses whose apices project into little bays of the pelvis called the *calyces*. The microscopic renal tubules, in which the urine is formed, discharge at the tips of the pyramids into the pelvis. The pelvis itself is partly enclosed within the kidney but protrudes at the hilum to become continuous with the ureter.

The kidney has a glistening true capsule, and lies embedded in a

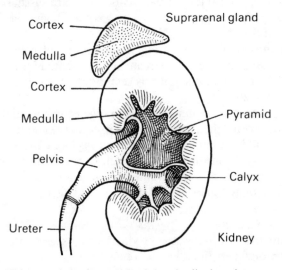

Fig. 8.17   Kidney and suprarenal gland, longitudinal section

voluminous false capsule of perinephric fat in which it glides up and down with respiration. Since the kidneys lie on the lower ribs, the lowest part of the pleural cavity of the chest overhangs the back of the upper pole, separated by the lowest fibres of the diaphragm; but most of the organ lies on the psoas and quadratus muscles. The other relations differ on the two sides.

Right: The right lobe of the liver is in front of most of the anterior surface; the second part of the duodenum lies before the hilum; the hepatic flexure of the colon is in front of the lower pole.
Left: The body of the pancreas crosses the middle of the anterior surface, which also forms part of the stomach bed. The spleen is applied to the convex outer border, and the splenic flexure of the colon is in contact with the lower pole.

The *ureters* begin at the pelviureteric junction in the hilum, emerge from behind the renal vessels, and run down and in on the psoas muscle behind the peritoneum. At the pelvic brim, they cross in front of the common iliac vessels, run down the side wall of the true pelvis and enter the bladder. They are hollow muscular tubes, some 25 cm long and 4 mm in diameter, down which urine is propelled in spurts by waves of contraction.

The diaphragm
This is the chief muscle of respiration and forms a partition, part muscular, part tendinous, between thoracic and abdominal cavities. It has a right and left muscular *dome* rising high into the thorax and separating the lungs from the abdominal viscera; and an intermediate flat *central tendon*, on which rests the heart.

The muscular portion has a number of bony origins. In *front*, it arises from the back of the xiphisternum; at the *sides*, from the lower ribs; and in the midline *posteriorly* it helps form the upper part of the posterior abdominal wall by arising in two pillars or *crura* from the sides of the upper three lumbar vertebrae (Fig. 8.18).

There are several apertures for structures passing between thorax and abdomen. The *aortic aperture* is embraced by the two crura as they cross in front of the twelfth thoracic vertebra. The *oesophageal aperture* is in the left dome at the level of the tenth thoracic body, and the opening for the *inferior vena cava* lies in the central tendon to the right of the midline at the level of the ninth thoracic body.

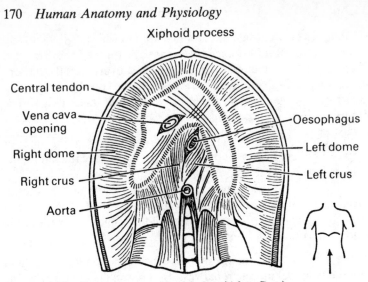

Fig. 8.18   The diaphragm, seen from below (After *Gray*)

The pelvis

The lower portion of the abdominal cavity is known as the pelvic cavity and is set inclined backward to form rather a separate compartment, although the two are continuous at the pelvic inlet (Fig. 8.1).

The cavity is formed by the basin-shaped arrangement of the bony pelvis, formed by the two innominate bones at the side and in front, and the sacrum and coccyx behind (Figs 3.13, 3.14). The bony pelvis falls in two parts (Figs 8.20, 8.23):

1   The *false pelvis*, the shallow edges of the basin above the pelvic brim, i.e. the right and left *iliac fossae*, clothed by the iliacus muscles and supporting the caecum and pelvic colon respectively;

2   The *true pelvis*, the deeply enclosed portion below the pelvic brim containing the essential pelvic organs.

The true pelvis is what is usually meant by the word, and the following remarks apply to it alone. It has an *inlet*, an *outlet*, a *cavity* and a *floor*. Its boundaries are only partly bony and completed by soft tissues, and the floor is made up entirely of soft tissues.

The *inlet* faces forwards and slightly upwards, owing to the backward inclination of the sacrum. Its boundary is the pelvic brim, seen from above to be heart-shaped, with the projecting upper part of the sacrum, the promontory, encroaching behind and the angle of the pubic arch meeting at the symphysis in front (Figs 3.13, 3.14).

The *cavity* is a conical canal. Its walls are: *anteriorly*, the back of the pubic symphysis and rami; *laterally*, the innominate bones, with the obturator foramina bridged by the obturator membranes; *posteriorly*, the anterior surface of sacrum and coccyx. These bony walls are incomplete and are lined by muscles, pelvic fascia and peritoneum. The principal contents are the pelvic colon and rectum on the posterior wall, in the hollow of the sacrum, and the bladder in front, behind the symphysis. In women the uterus and vagina are interposed between rectum and bladder; in men the bladder rests on the prostate gland and seminal vesicles.

The *outlet* is a diamond-shaped space seen from below. The four bony points of the ◇ are: anteriorly, the symphysis; at each side, the ischial tuberosity; and posteriorly, the tip of the coccyx. The two anterior limbs of the ◇ are bony, the pubic arch formed by the inferior pubic rami running up to the symphysis; the two posterior limbs are ligamentous, the sacrotuberous ligaments spanning from sacrum and coccyx to ischial tuberosities.

Sex difference in the bony pelvis

As the female pelvis has to allow the passage of the baby's head, it is roomier than the male pelvis and shallower. The female pelvis has its side walls more vertical; the iliac fossae are shallower; the inlet is large and nearly circular; the sacrum is short and wide, and only projects a little into the cavity; the outlet is wide, the pubic arch forming an obtuse angle, while the coccyx is very mobile.

The male pelvis is a narrow cavity with sloping walls, a small heart-shaped inlet with marked sacral encroachment, a tighter outlet with the pubic arch an acute angle, and a more rigid coccyx.

Surface anatomy

The only region where the pelvis comes near the surface is at the *perineum*, the space between the legs which contains the orifices of genital, urinary and digestive tracts. Fig. 8.19 shows the female perineal structures. There is a diamond-shaped area of perineal skin

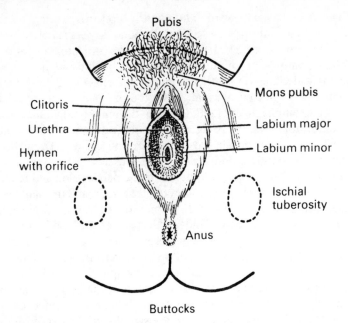

Fig. 8.19   The female perineum

divided into two triangles by a line joining the ischial tuberosities. The posterior *anal triangle* contains the opening of the anus, continuous with the rectum above. The anterior *urogenital triangle* contains the opening of the *vagina*, the lower end of the genital tract, with the *urethra* just in front, the urinary channel which runs down from the bladder through the pelvic floor; further forward is the *clitoris*, the diminutive female equivalent of the masculine penis.

In *men* the anal triangle is similar, but the urogenital region is different. The penis has at its base a central *bulb*, attached to the midpoint of the perineum, into which the urethra passes after its passage from the bladder through the prostate and pelvic floor (Fig. 8.21); and it has two supporting *crura*, attached to the pubic arch on each side and converging on the bulb to form the shaft of the organ, which is traversed by the *urethral canal*, opening at the *external urinary meatus* at its tip.

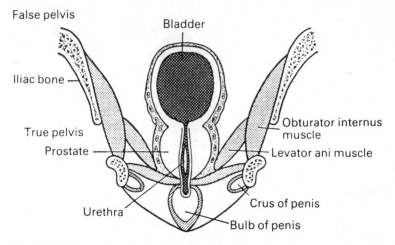

Fig. 8.20 Coronal section of male pelvis. Note the distinction between true and false pelvis; also the muscles of the pelvic floor (After *Gray*)

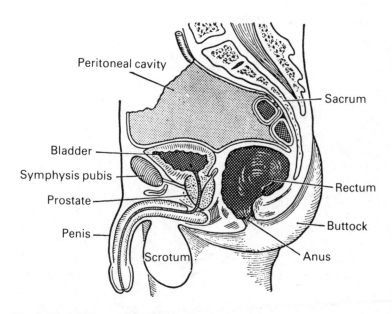

Fig. 8.21 Midline sagittal section of the male pelvis (After *Gray*)

The pelvic floor

The pelvis is like a funnel with a very wide stem; the false pelvis forms the sloping sides of the funnel and the true pelvis to the vertical portion. The pelvic organs are contained in the vertical portion and rest on its floor, a muscular partition slung from the sidewalls of the cavity. This embraces the various canals – anal, vaginal, urethral – which pierce it to reach their perineal orifices, surrounding each with a sphincteric grip. The upper aspect of the pelvic floor is lined by the parietal layer of peritoneum, which is reflected over the contained organs. The general arrangement is shown in longitudinal section (Figs 8.21, 8.22) and is considerably simpler in the male.

In the *male* the peritoneum is reflected from the back of the abdominal wall over the upper surface of the bladder, dips down between bladder and rectum (the rectovesical pouch) and turns up again over the front of the rectum. Most of the bladder is below peritoneal level, the prostate entirely so.

In the *female* the peritoneum, after covering the bladder, is thrust up by the projection of the uterus; it covers the front and back of that organ, passing off the latter to form the recto-uterine pouch –

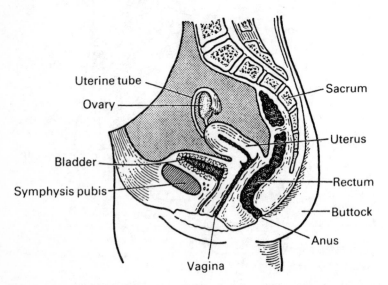

Fig. 8.22   Midline sagittal section of the female pelvis

the most dependent part of the abdominal cavity – before ascending again on the rectum.

In both sexes the lower limit of peritoneum is the third piece of the sacrum; the upper part of the rectum has a peritoneal covering in front, but no mesentery, the lower part is entirely below peritoneal level.

Pelvic viscera

Bladder and rectum are the same in both sexes; the genital organs will be dealt with in Chapter 21.

Bladder. This is a hollow, muscular organ in the front of the pelvic cavity; it receives urine via the ureters and expels it into the urethra by micturition. Its shape varies with distension, but is roughly that of an inverted pyramid (see Fig. 21.2). When empty, it lies entirely within the pelvis. As it distends, its domed upper portion or *fundus* ascends into the abdomen, in contact with the back of the posterior abdominal wall.

Its upper surface is covered by peritoneum, and the ureters open into its upper lateral angles by oblique passages through the muscular wall, at the *uretic orifices*. The tapering dependent bladder neck is continuous with the urethra at the *internal urinary meatus*. The two ureteric orifices and the internal meatus form the three points of the *trigone*, a region of the bladder base, sensitive to stimulation.

In *women* the organ is overhung from behind by the uterus, and the urethra is a canal of only 4 cm which pierces the pelvic floor to open immediately in front of the vagina. In *men* the neck of the bladder rests on the *prostate gland*, into which the male urethra passes; the *seminal vesicles* for storage of sperm, and the ducts conveying sperm from the testicles, are close to the lower part of the male bladder.

The bladder wall has a thick muscular coat and a mucosal lining, which is wrinkled when the organ is contracted. The involuntary nervous system adjusts muscle tone to fluid content so that it relaxes as it fills, keeping the tension constant until a threshold is reached, when the tension is felt as an urge to micturate (see also p. 300).

Rectum and anus. The rectum is the lower part of the large bowel; faeces enter from the pelvic colon and remain until discharged by defaecation. It is some 13 cm long, begins as the continuation of the

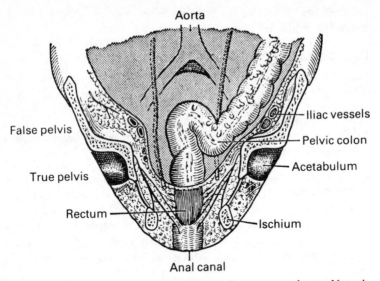

Fig. 8.23   Coronal section of pelvis to show the rectum and anus. Note the distinction between true and false pelvis; also the lower limit of the peritoneal cul-de-sac (After *Gray*)

pelvic colon at the middle of the sacrum, follows to the hollow of the sacral curve, and turns forward at the coccyx to join the *anal canal* – a short wide passage 4 cm long that bends sharply backward to open at the anal orifice.

The rectum has a wavy course, and its lower portion, or *ampulla*, is capable of considerable distension; two or three valve-like folds of mucosa project into its lumen. Only the upper portion has a peritoneal coat, on the front and sides of the bowel; its wall is longitudinal and circular muscle, and the lowest part is thickened as a powerful ring gripping the ano-rectal junction, the *internal sphincter*.

The anal canal is distensible for the passage of faeces; its upper part is lined with mucosa, but the lower 1·3 cm by inturned skin continuous with that of the perineum. Immediately under the peri-anal skin, encircling the orifice, is another muscle, the *external sphincter*. Both sphincters are normally closed; in the act of defaecation they relax, the abdominal wall tightens and raises abdominal pressure, and the rectum is lifted up over its contained faeces by the muscular pelvic floor.

Other pelvic structures

The *ureters* cross the pelvic brim by passing over the common iliac vessels; travel down the sidewalls of the pelvis under the peritoneum, 5 cm from the rectum; and sweep inward and forward to reach the bladder.

Vessels. The *common iliac* vessels pass downwards and out from their origin, and divide at the level of the lumbo-sacral joint into external and internal iliac branches. The *external iliacs* continue along the pelvic brim, together with the psoas tendon and the femoral nerve, and pass beneath the inguinal ligament to emerge as the femoral vessels in the groin of the thigh. The *internal iliac* vessels descend the sidewall of the true pelvis and branch to supply bladder, rectum and genital organs.

Nerves. The *lumbar plexus* on the posterior abdominal wall is succeeded in the pelvis by the *sacral plexus* of nerve roots, emerging from the foramina of the sacrum and lying on the back of the cavity. The main branch of the lumbar plexus is the *femoral nerve*, which accompanies the psoas tendon and external iliac vessels to the thigh; the main branch of the sacral plexus is the *sciatic nerve*, which leaves the back of the pelvis to emerge in the buttock.

The *sympathetic chain* continues down from the abdomen on the front of the sacrum and coccyx, ending as a single fused ganglion on the latter.

Uterus, uterine tubes, ovaries, vagina, prostate, seminal vesicles, testicles and their ducts are described in detail in Chapter 22.

9

Regional Anatomy: the Thorax

The thorax is a bony cage for the heart and lungs. It appears to be cylindrical but when the shoulders are removed (Fig. 9.1), it is seen to be conical or barrel-shaped, with a narrow apex.

Bony wall

In the midline behind are the twelve thoracic vertebrae and in the midline anteriorly the breastbone or sternum; between them are the encircling ribs. Note, in cross-section (Fig. 9.4), the paravertebral recess on each side of the spine due to the backward curve of the ribs before they turn forward.

Ribs

There are twelve pairs of ribs, separated by *intercostal spaces* which contain the intercostal muscles, nerves and vessels. The first seven are *true ribs*, complete from spine to sternum; the eighth, ninth and tenth are *false ribs*, turning up to join the ribs above; the eleventh and twelfth are short *floating ribs*, embedded in the flank muscles of the abdomen. The first and twelfth ribs are very short; the longest the seventh and eighth, so that the barrel contour of the chest slopes in again above and below. Each rib also slopes somewhat downwards.

Each rib has a *head*, articulating with the side of a vertebral body; a short *neck*, lying on the transverse process of the vertebra; a *body*, which runs back a little and then sweeps sharply forward at the *angle* forward; and, the *costal cartilage*, a gristly rod of 2·5 or 5 cm

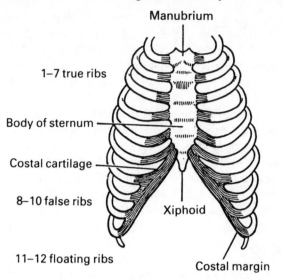

Fig. 9.1 The bony thoracic cage, seen from in front

connecting rib to sternum. The framework is deficient below anteriorly, where the costal arch is formed by the right and left costal margins.

The *intercostal spaces* are filled by layers of muscle, between which the intercostal vessels and nerves encircle the chest. One nerve, artery and vein occupy each space, a good example of segmentation (p. 69).

The *sternum* or breast bone is dagger-shaped, with three main

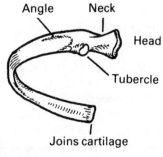

Fig. 9.2 A typical rib, seen from behind

pieces. Uppermost is the broad, flat *manubrium*, to which are attached the inner ends of the clavicles and first ribs. Below is the main *body*, a flat bone of two compact layers enclosing spongy bone and red marrow, with the costal cartilages attached on either side. Lowest is the small pointed *xiphisternum*, in the upper abdominal wall.

The thorax has an *inlet* above, through which the great vessels and nerves pass into the neck or over the first rib into the axilla and arm. The inlet is narrow and closely packed, only 6·5 cm separating the manubrium from the spine, with the first ribs on either side. The most important structures traversing the inlet are the *oesophagus* (gullet), *trachea* (windpipe), certain *nerves*, and the great arteries and veins. The thorax is sealed off below by the diaphragm.

Respiration

Respiration draws air into and pushes air out of the lungs. In inspiration the chest cavity is enlarged and air enters; in expiration the reverse occurs. Respiratory movements are both thoracic and abdominal. In *thoracic* inspiration the sternum is lifted by elevation of the ribs, which come to lie more horizontally; both the transverse and anteroposterior diameters of the chest are increased. In *abdominal* inspiration the diaphragm contracts and flattens, pressing the abdominal organs down and bulging the abdominal wall; the vertical height of the thorax is increased. Inspiration is an active process, due to muscular exertion. Expiration is passive; the chest wall subsides, the abdominal wall recoils, the diaphragm relaxes and air is driven out of the lungs.

The muscles of the chest wall

Anterior

In front, the chest wall is covered by the great *pectoralis major*, referred to in connection with the arm. Arising from sternum, clavicle and ribs, its fibres converge to be inserted into the upper humerus; its function is adduction of the arm to the side. A smaller *pectoralis minor* lies under cover of the major, arising from a few ribs and inserted into the coracoid process of the scapula. In the lower chest wall, some of the abdominal muscles are attached to the ribs – the *rectus* near the midline and the *oblique* muscles more laterally.

Posterior (Fig. 9.3)

The two great superficial sheets of muscle at the back are the trapezius (above) and latissimus (below). The *trapezius* is triangular, with its base attached to the spinous processes of the thoracic vertebrae and extending up the back of the neck as far as the occipital region of the skull. The muscle narrows rapidly to its insertion on the spine of the scapula and the back of the clavicle. It

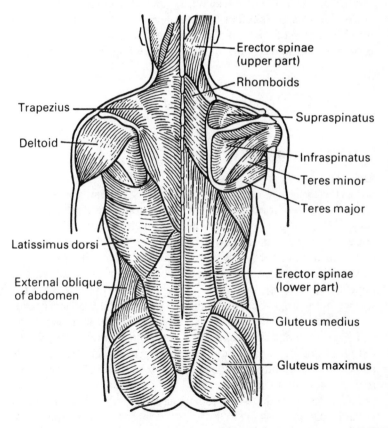

Fig. 9.3   Muscles of the back. On the right side the superficial muscle layer has been removed to expose the deeper structures

rotates the scapula on the chest wall (Fig. 6.10) and thus aids abduction at the shoulder joint.

The *latissimus dorsi* is overlapped by the trapezius above. It originates from the lower thoracic vertebrae and lumbar spine, via a dense sheet of fascia. Its fibres pass away and up to be inserted into the humerus at the same level as the pectoralis major; it adducts the arm. With fingers in the armpit, a thick muscle boundary is felt in front and behind; in front is the pectoralis; behind is the latissimus.

Under cover of the superficial muscles is the *erector spinae* muscle, which forms a thick, rounded belly stretching from skull to sacrum on each side of the spine. It gains attachment to the ilium, vertebrae, ribs and skull at different levels, and helps maintain the normal curves of the spine in the erect posture. Its action is to extend, i.e. straighten out, the spine – including the head and neck.

Surface anatomy

Anterior

If the clavicles are traced to their medial ends, the manubrium of the sternum can be felt between them, in the hollow of the base of the neck. Five cm lower, in the midline, the junction of manubrium and body of sternum is marked by a prominent angle; it is easy to feel the costal cartilage of the second rib attached here on either side.

The main body of the sternum is subcutaneous, a broad plate of bone ending below as the little xiphoid process, depressed below the surface in the angle between right and left costal margins. The ribs are attached on each side of the sternum and are best numbered by locating the second rib, as indicated, and then counting downwards; the inner end of the first rib lies too deep to be felt. The ribs are palpable well out into the side and the intercostal spaces can be felt as depressions between.

The *nipple* lies between fourth and fifth ribs in men, but is variable in women owing to the size of the breast. The beating of the apex of the heart is felt or seen just medial to the left nipple, i.e. in the fifth space on the left side, some 9 cm from the midline.

Note that the lungs do not extend quite as far as the lower borders of the thoracic cavity. This does not imply an empty pleural space below; there is simply a narrow cleft between diaphragm and chest wall where the two pleural layers come in contact (see page 184).

Posterior

The back of the chest is obscured by the overhanging shoulder blades and the great muscle masses of this region. The spinous processes of the thoracic vertebrae are felt in the midline, but the ribs lie too deeply to be felt easily. The eleventh and twelfth ribs are short, extending for only a few inches on each side of the spine.

Thoracic cavity

The thoracic cavity is divided into right and left halves by a massive partition, the *mediastinum*, which lies in the sagittal midplane, stretching from the back of the sternum to the vertebral column. Vertically, it extends from thoracic inlet to diaphragm. The two halves are separate, and contain the right and left lungs. Each lung is

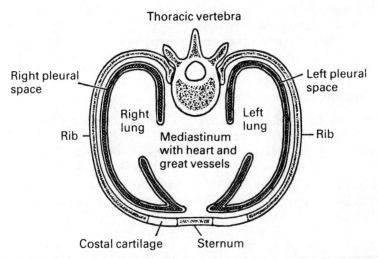

Fig. 9.4   Cross-section of the chest to show its division into right and left halves by the mediastinal partition

attached to the mediastinum at its root or *hilum* by its bronchus (air-tube) and blood-vessels.

The cavity is lined by a smooth serous membrane, the *pleura*, which facilitates gliding of the lung on the chest wall. Each pleural space is a closed sac, for the pleura, like the peritoneum, is divided into *parietal* and *visceral* layers which are everywhere continuous. The *parietal pleura* clothes the deep aspect of the ribs, the upper surface of the diaphragm and the sides of the mediastinum; the *visceral layer* encloses the lung, the two layers being continuous along its root of that organ. The apex of the pleural sac rises into the root of the neck, a little above the clavicle, and its base overhangs the liver in front and the kidneys behind. Although the lung closely follows the pleura, it does not quite reach its upper and lower limits. Normally there is no actual pleural cavity, for the two layers are in contact, the lung filling its side of the thorax. But the lung is very elastic and tends to shrink and expel its contained air. This cannot occur normally, and a negative pressure exists in the potential pleural space; if air is admitted, the lung immediately shrinks into a small solid mass, collapsed against the mediastinum.

The *heart* is embedded in the mediastinum, occupying a central position between the two lungs, though extending more to the left. It is enclosed in a fibrous sac, the *pericardium*, which rests on the central tendon of the diaphragm. After removal of the lungs; it can be seen how the intervening mediastinal partition is packed solidly with vital structures – heart, great vessels, trachea and oesophagus.

Contents of the thoracic cavity

The general arrangement of heart and lungs is indicated in Figs 5.8, 9.5. Fig. 5.8 shows the thoracic viscera after removal of the anterior chest wall, and how the lungs overlap the heart so that only a small part is in contact with the back of the sternum. Fig. 9.5 shows the organs removed from the chest, with the lungs turned back to expose the great vessels and heart; the pericardium enclosing the latter has been opened.

Trachea and bronchi (Fig. 5.10)
The *trachea* enters the chest from the neck at the thoracic inlet and runs down to the sternal angle, where it divides into a *right* and *left*

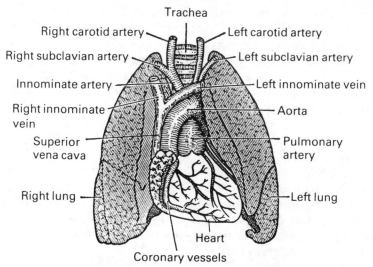

Fig. 9.5   Lungs, heart and great vessels, seen from in front after removal from the chest. The lungs normally overlap the greater part of the heart and have been retracted to expose that organ (After *Gray*)

main *bronchus* for each lung. In its short course it has the oesophagus behind, separating it from the spine, and the great vessels – aorta and superior vena cava – in front, separating it from the back of the sternum. The bronchi enter the lung roots, together with the pulmonary vessels. All these structures emerge from the mediastinum to reach the hilum of the lungs. Trachea and bronchi have numerous cartilaginous rings, which keep them from collapsing and maintain a free air passage. In the trachea these rings are deficient behind, so that the windpipe is semicircular in section (see Fig. 10.8).

Lungs

The lungs are light, spongy organs designed to provide the maximum surface area for the interchange of oxygen and carbon dioxide between air and blood. Each lung is divided into a main upper and lower lobe by an oblique fissure running downwards and forwards; the right lung has a third lobe, marked off from the upper lobe by a short transverse fissure. The lungs are permeated by elastic tissue; when removed from the chest, they collapse to a quarter of their normal size. The whole or part of the organ floats in water.

The lungs lie free in the chest, enclosed in the visceral pleura, attached only by their roots to the mediastinum. Each *apex* rises into the root of the neck 1·3 cm above the middle of the clavicle. The *base* rests on the upper surface of the diaphragm, which separates it on the right from the right lobe of the liver and on the left from the left lobe, stomach and spleen. The *outer surface* is in contact with the inner aspect of the chest wall and is marked by the ribs. The rounded *posterior border* fits into the recess at the side of the thoracic vertebrae, and the *anterior border* overlaps the heart and pericardium. Because of the bulk of the heart on the left side, there is a well-marked *cardiac notch* in the anterior border of the left lung.

The *medial surface* faces inwards towards the mediastinum and on it is situated the hilum. The relations of the medial surfaces differ on the two sides, for the mediastinal structures are not symmetrical. The main differences are:

1   The curvature of the arch of the aorta over the left bronchus to the left side and the prominence of the descending aorta on the left side;
2   The presence of the superior vena cava on the right;
3   The right-sided position of the oesophagus below.

Internal structure of the lungs
Each main bronchus divides into a branch for each lobe, and these branches subdivide into a ramifying tree of fine *bronchioles*, accompanied by branches of the pulmonary artery and vein. Each bronchiole ends in a cluster of tiny air-filled sacs or *alveoli*, and interchange occurs between their contained air and gases dissolved in the blood of the capillaries around the alveoli. Each lung lobule is composed of a terminal bronchiole and its air cells, and the lung has millions of lobules bound together by elastic connective tissue.

Heart
The heart is a hollow, muscular organ lying between the lungs in the mediastinum; it is enclosed in the *pericardium*. This membrane has a tough outer *fibrous layer*, firmly blended below with the central tendon of the diaphragm, and a delicate inner *serous layer*. The serous pericardium lines the fibrous capsule and is reflected at the roots of the great vessels of the heart to cover that organ, i.e.

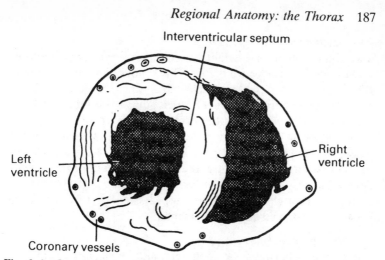

Fig. 9.6 Cross-section of heart in ventricular region. Note the greater thickness of the left ventricle, the massive septum and the coronary vessels cut across.

there are parietal and visceral layers just as with the pleura and peritoneum, with a potential pericardial cavity.

The heart has four hollow chambers: the two *atria* above, which receive blood from the great veins, and the two *ventricles* below, which expel it into the great arteries. Although the atrium and ventricle of the same side communicate freely, they are shut off from the opposite pair by a muscular *septum*, which is much thicker between the ventricles than between the atria. The latter are thin-walled compared with the fleshy ventricles, and the wall of the left ventricle is thicker than the right. On the surface, grooves indicate the demarcation between the chambers, and in these grooves run the important *coronary vessels* which nourish the muscle of the heart (Fig. 9.5).

The shape of the heart is roughly conical, with the apex directed to the left and slightly downwards. The anterior surface – behind the sternum and costal cartilages – is formed mainly by the right ventricle with a little of right atrium and left ventricle on either side. The right border is entirely right atrium, the left border is left ventricle and the apex is the tip of the latter. The left atrium lies entirely on the posterior surface, facing the vertebral column, oesophagus and descending aorta.

The attachments of the great vessels

Two great veins enter the right atrium: the *superior vena cava* above and the *inferior vena cava* below. This venous blood flows through the chamber to the right ventricle and is expelled into the *pulmonary artery* arising from its upper border, going to the lungs in the two branches of this vessel. At the back of the heart, four *pulmonary veins* return fresh blood from the lungs to the left atrium, which passes it to the left ventricle; the latter expels it to the great *aorta*, the wide arterial trunk that is the commencement of the general circulation. Although the aorta rises on the left and the pulmonary artery on the right, this relation appears reversed from in front because the two vessels are intertwined.

The interior of the heart

The openings between the chambers, and between the ventricles and their great arteries, are guarded by a system of *valves*. Between

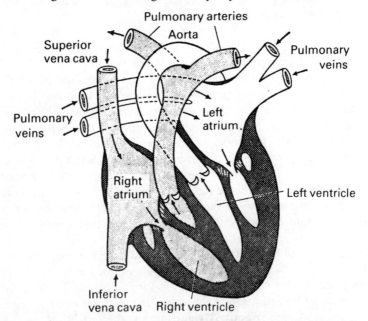

Fig. 9.7 Diagram showing the interior of the heart, entry and exit of the great vessels, and the direction of blood flow. The diagram is artificial in that no section normally exposes all four chambers in this way

each atrium and its ventricle is a parachute-like valve, the strands of which are attached to the ventricular wall, the flaps lying in the orifice between the chambers. On the right this is the *tricuspid* valve, with three flaps, on the left the *mitral*, with two flaps. When the atria contract, the bloodstream easily separates the flaps as it passes into the ventricles; but when the latter contract in *systole*, the flaps are ballooned out, the guy-ropes are tautened and no backflow is possible, all the blood being directed to the arterial exits. These latter, the openings into the pulmonary artery and aorta from the right and left ventricles, are guarded by simpler valves composed of three semilunar pockets facing upwards. These are easily pushed apart by the blood, but fall together when the ventricles relax in *diastole*, the weight of the blood column in each pocket forcing them into contact and blocking any reflux.

Great vessels of the thorax

The *superior vena cava* is formed by the junction of the right and left *innominate veins*, each bringing venous blood from one arm and one side of the head and neck. The right innominate and the vena cava run vertically, continuous with the right atrium. But the left innominate crosses from left to right in front of the trachea and above the arch of the aorta.

The *inferior vena cava* rises from the abdomen through the central tendon of the diaphragm and has only a very short intrathoracic course before it enters the lower part of the right atrium.

The *pulmonary artery* arises at the upper border of the heart from the summit of the left ventricle and twines round the aorta. Behind the aortic arch, it divides into right and left branches, which enter the roots of the lungs with the pulmonary veins and bronchi. The *pulmonary veins* are not seen from in front; two on each side, they drain into the back of the left atrium.

Aorta

This is the largest artery of the body, the beginning of the general or *systemic* circulation – as opposed to the local *pulmonary* circulation. It arises at the upper border of the heart from the summit of the left ventricle and its subsequent course is as follows:

1   A short *ascending portion*, running up to the left of the superior vena cava;

2   The *aortic arch*, curving horizontally backwards in front of the
     lower part of the trachea and over the left bronchus;
3   The *descending aorta*, running down on the vertebral column at
     the back of the thoracic cavity to pass through the diaphragm,
     where it is continuous with the abdominal aorta.

Branches of the aorta
From the summit of the aortic arch three great vessels arise. On the
right is the *innominate*, which divides into the right *subclavian*
artery for the arm and the right *common carotid* for the head and
neck. On the left there is no innominate artery, the left subclavian

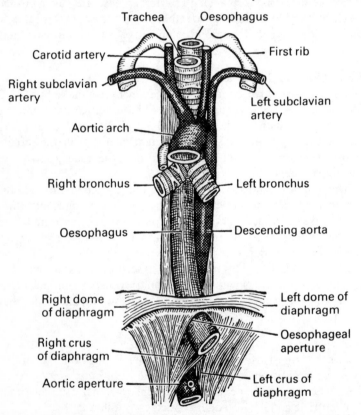

Fig. 9.8   Structures of the posterior thoracic wall: oesophagus and des-
          cending aorta. The heart and lungs have been removed (After
          Gray)

and carotid springing directly from the arch. Only the lower parts of these vessels lie within the thorax, the level of the innominate bifurcation being at the inner end of the clavicle; they are closely applied to the trachea. The corresponding veins lie more superficially and there is an innominate vein on both sides; the vessel equivalent to the common carotid artery is the internal jugular vein. The little *coronary arteries* arise from the very beginning of the aorta, immediately above the semilunar valves. There is a right and a left vessel, and any disease of these impairs the circulation in the muscular wall of the heart.

The *descending aorta* gives off on each side the pairs of *intercostal arteries* that encircle the chest in the intercostal spaces; the corresponding veins do not enter the venae cavae directly but through intervening channels, the *azygos* veins.

The aortic arch forms a great sweep from the front of the chest to the back, with an upward convexity from which arise the main arteries of the head, neck and arms. The concavity of the arch embraces the pulmonary arteries and veins and the bronchi, entering the roots of the lungs. There is also a little artery running the length of the back of the sternum, from above downwards. This is the *internal mammary* artery, a branch of the subclavian in the neck; it ends below by entering the sheath of the rectus muscle as the superior epigastric artery to anastomose with the inferior epigastric from below.

The arch of the aorta inclines to the left, so that the descending portion is to the left of the midline and nearer the left lung.

Remaining structures of the posterior thoracic wall

The *oesophagus*, or gullet, enters the chest from the neck at the thoracic inlet, lying immediately in front of the vertebral column; it travels down the back of the cavity between the roots of the lungs to traverse the diaphragm. It inclines forward and to the left in its passage, allowing the descending aorta to intervene between its lower portion and the spine; and it is more closely related to the right than the left lung. It is a hollow, muscular tube down which food is propelled to the stomach by waves of contraction or *peristalsis*; unlike other parts of the bowel, it has no serous coat as it lies entirely away from the pleura at the back of the mediastinum.

The *thoracic duct* begins below as the upward continuation, through the aortic aperture of the diaphragm, of the cisterna chyli of the abdomen, the channel receiving digested fat from the bowel and the lymphatic drainage from the legs. The duct ascends on the front of the vertebral column in the chest, lying to the right of the aorta, behind the oesophagus; it crosses to the left side at the level of the aortic arch, enters the root of the neck on that side, and discharges into the junction of the left subclavian and internal jugular veins.

Nerves within the thorax
The important nerve trunks in the thoracic cavity are:

1   The *vagus* nerves, one on each side. These are the tenth pair of cranial nerves which have travelled down the neck via the thoracic inlet. The *right* vagus runs alongside the trachea, passes behind the right lung root, where it breaks up into a pulmonary plexus, reforms as a single trunk and accompanies the oesophagus through the diaphragm. The *left* vagus runs with the left carotid and subclavian arteries, crosses the arch of the aorta, passes behind the left lung root to form a similar plexus and finishes its thoracic course alongside the oesophagus. The vagi are part of the *parasympathetic* system described on pp. 325, 327 concerned with automatic regulation of the viscera; in the chest they give branches to heart, lungs and oesophagus.

2   The *sympathetic* trunks lie one on each side, 3–5 cm away from the vertebral bodies. They consist of a chain of *ganglia* lying on the necks of the ribs behind the parietal pleura and connected by longitudinal fibres. This chain is continuous with that in the neck above and leaves the thorax below with the psoas muscle to become the abdominal sympathetic. It also is part of the *autonomic nervous system* and shares with the vagus the dual control of the viscera. Several branches run down from the main trunk; these are the *splanchnic nerves*, which pierce the diaphragm to join the coeliac or solar plexus surrounding the beginning of the abdominal aorta.

3   The *intercostal nerves*, derived from the spinal cord, run round the chest wall in the intercostal spaces. At their origins they lie deeply behind the pleura and are connected to the sympathetic ganglia by communicating twigs.

4   The *phrenic nerves* arise in the lower part of the neck to supply the diaphragm. The right phrenic runs down alongside the superior vena cava and the right side of the heart; the left crosses the aortic arch and the left side of the heart. Both pierce the diaphragm and break up on its undersurface.

10

Regional Anatomy: Head and Neck

The head

The skull

The skull is a jigsaw of bones, fitting into each other at the immobile fibrous sutures. Although complicated, it is easily understood by reducing it to its three essential parts: cranium, facial skeleton and mandible.

The *cranium* is the box enclosing the brain and its membranes; it has a domed vault and a base set deep beneath the soft structures.

The *facial skeleton* is attached to the underside of the base anteriorly, and includes the nasal bone, maxilla (upper jaw) and others.

The *mandible* or lower jaw is entirely separate and is slung to the underside of the base at the back.

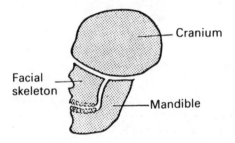

Fig. 10.1   The basic plan of the skull

The skull bones are mostly flat; in several places, particularly the vault, they have an outer and inner table of compact bone sandwiching a red-marrow-filled layer – the *diploë*. At certain sites they are expanded by internal air spaces or *sinuses*.

The vault is made up of four main bones: frontal, two parietals and occipital. The *frontal* is the substance of the forehead and overhangs the orbital cavities in front. The *parietals* lie on each side above the temples and the *occipital* at the back. The *coronal suture* separates frontal and parietals; the *sagittal suture* divides the two parietals in the midline; the *lambdoid suture* is the meeting-place of parietals and occipital. At birth, the angle of junction at either end of the sagittal suture is unclosed by bone and filled in with membrane, the anterior and posterior *fontanelles*; the posterior is closed by the end of six months, the anterior not till the second year.

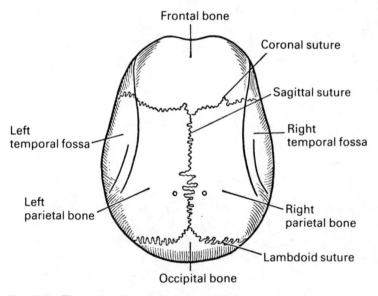

Fig. 10.2   The vault of the skull, or vertex, from above

Figs 10.3 and 10.4 are anterior and lateral views of the whole skull. Note, in the anterior view, the frontal bone coming down as a roof over the *orbital cavities*, the bony eyesockets; the openings of the *nasal* cavities, with the nasal bones above and the bony nasal

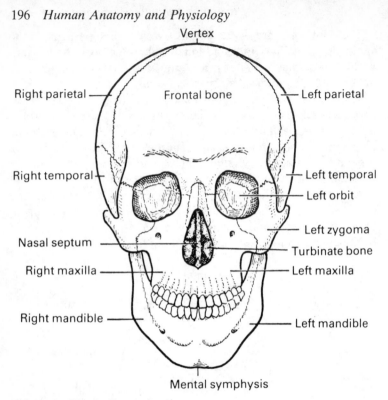

Fig. 10.3 The skull, anterior view

septum separating the two sides; and the *maxilla*, or upper jaw, carrying the upper teeth. Note, in the *lateral* view, the side aspect of the vault and how the temporal bone helps to form the lower part of the sidewall in the hollow above the cheek; the cheek bone, or *zygoma*, arching across from maxilla to temporal; the *mastoid process* behind the bony external opening of the ear canal; and the *styloid process* jutting out from the base below.

The skull base
Fig. 10.5 shows the upper surface of the skull base, after the skull cap has been removed. Note the great *foramen magnum* nearer the back, where the spinal cord is continuous with the medulla of the hindbrain; and the numerous smaller *foramina*, where veins and cranial nerves exit and the cerebral arteries enter. The inner table is

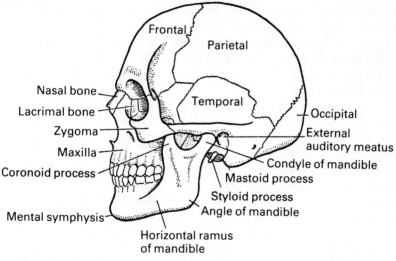

Fig. 10.4 Lateral view of skull

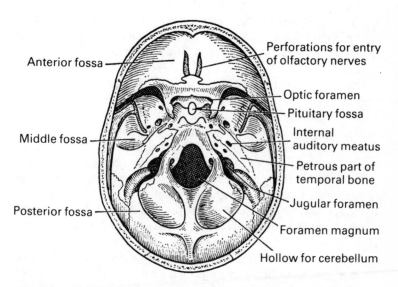

Fig. 10.5 The skull base, seen from within, after removal of the skull cap

grooved by narrow channels for the arteries supplying the brain membranes, or by wider paths for the venous sinuses which flow between the layers of the dura mater, the outermost covering of the brain. The base is divided on each side into three pockets or *fossae*, which become deeper and more extensive from before backwards.

The *anterior fossa*, in front, houses the frontal lobe of the brain; its floor is the roof of the orbit and nose. Where the two anterior fossae meet in the midline, tiny perforations exist for passage of the olfactory nerves from nose to brain.

The *middle fossa* houses the temporal lobe of the brain; through a little opening just behind the anterior fossa, the optic nerve connects eyeball and brain. In its floor are large foramina for the exit of the fifth or trigeminal cranial nerve – concerned with sensation in face and jaws, and movement of the masticatory muscles – and for entry of the internal carotid artery into the skull. Note, between the two middle fossae, a little pocket embedded in a small platform; this is the *pituitary fossa*, in which lies the pituitary gland.

The *posterior fossa* is separated from the middle by an oblique bony ridge, the petrous portion of the temporal bone, a mass enclosing the inner-ear cavity and the passage of the eighth (acoustic) nerve from ear to brain. The posterior fossa is mainly composed of the occipital bone; it contains the foramen magnum and a large foramen for the exit of the internal jugular vein, accompanied by various nerves.

On the undersurface of the base, one each side of the foramen magnum, are two articular processes or condyles, forming a joint with the first cervical vertebra, the atlas.

The *mandible* is formed by two halves joining in the midline, at the chin, the *mental symphysis*. Each half consists of a horizontal *body*, carrying the lower teeth on its alveolar (gum) margin, and a vertical *ramus*. Body and ramus are connected at the *angle*. At the upper end of the ramus are the pointed *coronoid process* and the rounded *condyle*, which articulates with the undersurface of the temporal bone at the *temporo-mandibular joint* in front of the ear (Fig. 10.4).

The *air sinuses* of the skull are accessory extensions of the nasal cavities, and their lining mucous membrane is continuous, though the connecting passages may be intricate. The main ones are the *frontal* and *maxillary*; there are also certain groups deep in the skull – *ethmoids* and *sphenoids*. The maxilla is entirely hollowed out by a great cavity, the *maxillary antrum*. The air cells which honeycomb the mastoid process are independent structures.

The *hyoid* is a little bone which can be felt in the neck in front, between chin and larynx, giving attachments to the muscles of the floor of the mouth and the tongue. It has a tiny body with spreading wings.

Surface anatomy of the head

Most of the external surfaces of the cranium, facial skeleton and mandible are easily palpable; the base is inaccessible. In the *cranium* the vault can easily be felt through the scalp, which is mobile over the bone. In front the overhanging ridges of the frontal bone, which are the upper margins of the orbital cavities, beneath the eyebrows, meet in the midline at the root of the nose. There are four bulges on the vault: the *frontal eminences* of the forehead and the *parietal eminences* above the ears. The lowest accessible part of the skull in the midline posteriorly is the external protuberance of the occipital bone, arrived at by tracing upwards the median furrow between the muscle masses at the back of the neck.

In the *face* the *nasal* bones brace the upper part of the nose; the bulbous tip is supported by cartilage only. A finger in the nostril touches the *septum* between the two cavities; this is cartilage in front, the bony septum being further back. The *zygomatic arch*, the cheek bone, is felt running between orbit and ear; above it is the depressed temporal fossa, in which lies the *temporalis* muscle which shuts the mouth and can be felt to contract on clenching the teeth.

The *auricle* is the external part of the ear, the aperture being the *external auditory meatus*; this channel is cartilaginous to begin with, its bony portion further in. Behind the ear is the mastoid process, with its tip below; just in front of the ear, below the zygoma, the condyle of the mandible can be felt moving at the *temporomandibular joint* when the mouth is opened and closed.

Body, ramus and angle of the mandible are all accessible, though the coronoid process is not palpable. The outer surface of the ramus is covered by the *masseter* muscle, which closes the mouth and stands out on biting. If this muscle is contracted by clenching the teeth, a small tubular structure can be rolled under the skin, parallel to the zygoma and a finger's breadth below; this is the duct of the parotid salivary gland running forward to open into the mouth.

The *scalp* includes all the structures overlying the vault. Its skin is profusely supplied with hair follicles, and the subcutaneous fat is thin and dense; beneath is a muscle sheet, which is most developed in the frontal and occipital regions as the frontalis and occipitalis bellies, connected by an aponeurotic sheet – the *galea* – stretching over the vault. The galea is separated from the bone by loose areolar tissue, and all the layers of the scalp move as one over the skull when its muscles contract, as in raising the eyebrows. The scalp veins communicate freely with those of the diploë, the venous sinuses of the dura and of the brain.

The face
The *facial skin* is thin, mobile and vascular, and the facial muscles lie immediately beneath (Fig. 10.6). Eyes and mouth are surrounded by circular muscles, the *orbicularis oculi* and *orbicularis oris*, which shut the eyelids and lips. Small muscles attached to the nasal cartilages wrinkle the nose and dilate the nostrils. Others lift or depress the corners of the mouth and the *platysma* of the neck is a broad sheet which spreads up over the mandible to share control of expression.

All these are supplied by the seventh (*facial*) cranial nerve, which leaves the base of the skull in front of the mastoid, runs forward in the parotid gland, and breaks up into branches to face and scalp at its anterior border.

The *muscles of mastication* are a separate deeper group, supplied by the fifth (*trigeminal*) cranial nerve. They include the *masseter* and *temporalis* – responsible for biting – and two *pterygoids*, lying deep on the inner side of the ramus, responsible for the chewing movements. A very deep muscle, the *buccinator*, runs transversely between mouth and masseter in the substance of the cheek.

The *parotid salivary gland* lies under the skin in front of the ear, covering the back of the masseter, and reaches up to the zygoma and down to the mandibular angle. The facial nerve is embedded in it, and it overlies the internal carotid artery and internal jugular vein as they travel between the neck and skull base. The *parotid duct* runs forward on the masseter to open into the mouth by piercing the buccinator muscle opposite the second molar tooth.

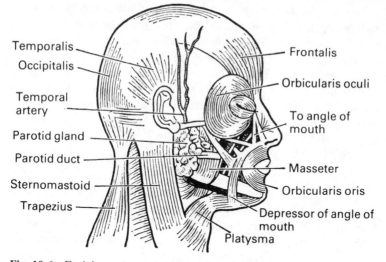

Fig. 10.6 Facial muscles and associated structures

The other external salivary gland is the submandibular, which lies below, and partly under cover of, the angle of the mandible; its duct pierces the floor of the mouth.

Facial vessels and nerves

The *superficial temporal artery* runs in front of the ear, where it can be felt pulsating, to the scalp. The *facial artery* ascends from the neck over the body of the mandible, obliquely upwards to the angle of eye and nose, giving branches to the lips. Both vessels are branches of the external carotid.

The seventh (facial) nerve is mainly motor to the muscles of expression.

The sensation of face and scalp is supplied by the fifth (trigeminal) nerve, which also innervates the masticatory muscles.

Most of the lymph drainage of the face is to groups of parotid and submandibular glands; that from the lower lip goes to small glands under the chin.

The mouth

This is the first part of the digestive tract. It is bounded by the cheeks and lips at the sides, roofed by the palate; and its floor is the muscular oral diaphragm, separating it from the neck, in which is embedded the root of the tongue. The oral cavity is lined by a glandular mucous membrane and is continuous behind with the pharynx. The *vestibule* is the space between lips and teeth.

The red surface of the *lips* is the mucous membrane, sharply demarcated from the skin at the mucocutaneous junction; each lip has a branch of the facial artery running parallel to its free margin, and felt pulsating when the lip is gripped between finger and thumb. Each is attached at its midpoint to the gum by a fold of mucous membrane.

The teeth

These are set in the opposed (alveolar) margins of the upper and lower jaws, where mucosa and periosteum are firmly blended to form the gum. The first, temporary, set of milk teeth erupt at intervals throughout the first two years; the second, permanent, set begin to replace them at six years and are complete by twenty-five, although the last molar or wisdom tooth is often delayed, or may never appear.

The teeth are named, from front to back, as *incisors, canines* (tearing), *premolars* and *molars* (grinding), and are symmetrical as between either side and between upper and lower jaws, so that each quarter of the whole set is the same. This may be expressed in the formulae:

	Molar	Premolar	Canine	Incisor	Incisor	Canine	Premolar	Molar	
Upper	3	2	1	2	2	1	2	3	= 32,
Lower	3	2	1	2	2	1	2	3	

for the permanent dentition, and:

Upper	2	0	1	2	2	1	0	2	= 20,
Lower	2	0	1	2	2	1	0	2	

for the milk dentition. Note the absence of premolars in the latter.

Each tooth has a *crown*, projecting beyond the gum; a *root*, embedded in the alveolus, and an intermediate *neck*. Longitudinal section shows the central *pulp cavity*, outside this the *dentine* or ivory, the bulk of the tooth, and an outer coating of hard *enamel*. The vessels and nerves enter the pulp through a small foramen at the tip of the root (Fig. 10.7).

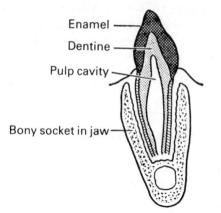

Fig. 10.7   Longitudinal section of tooth

The *palate* is the roof of the mouth and the floor of the nose. It consists of the bony *hard palate* in front and the mobile, muscular *soft palate* behind, projecting into the pharynx. The soft palate is arched, with its supporting pillars attached on either side of the back of the tongue; each pillar splits to enclose the *tonsil*, a mass of lymphoid tissue. From the free posterior edge of the soft palate there hangs down a small conical process, the uvula.

In the *floor of the mouth*, a fold or *frenulum* is seen tethering the underside of the tongue to the mucous membrane; the duct of the submandibular salivary gland opens on each side on a little papilla. Between the frenulum and the jaw on each side, a loose mucosal flap can be seen, or rolled by the tongue; this is the sublingual fold overlying the *sublingual salivary gland*, whose small ducts open directly into the mouth. Parotid, submaxillary and sublingual glands make up the three main salivary glands.

The *tongue* is a muscular organ concerned with swallowing, taste and speech. Its main anterior portion lies horizontally in the floor of the mouth; the back is curved round vertically and forms the anterior wall of part of the pharynx. The *root* of the tongue is embedded in the oral diaphragm; it is attached by various muscles to the hyoid bone and mandible, and to the soft palate by the arches of the latter.

The *dorsum* is the upper surface, with a number of *papillae*, hair-like processes, and *taste buds*. The tip rests against the back of the upper teeth and the undersurface is attached to the floor of the mouth by its frenulum. The organ is mainly muscular, with right and left halves separated by a fibrous partition; it has a mucous covering specialised for the sensation of taste.

The neck

The neck connects the head and trunk; its bony framework is the seven cervical vertebrae behind, and it is traversed by the food and air passages on their way to the chest, and the great vessels and nerves running between the thoracic inlet and the base of the skull.

At the root of the neck on each side is an outflow of nerves and vessels over the first ribs into the arms.

Cross-section of the neck shows the relative disposition of these structures. But whereas a section above the level of the sixth cervical vertebra shows the air and food passages as larynx and pharynx, at a lower level these have become the trachea and oesophagus.

The main points are (Fig. 10.8):

1 The supporting *cervical vertebrae*, nearer the back;
2 The thick *muscle masses* applied to the back of the spine;
3 The *oesophagus*, immediately in front of the spine;
4 The *trachea* between skin and oesophagus;

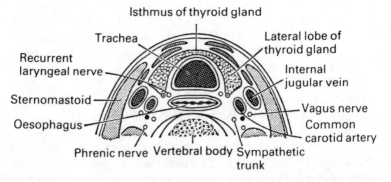

Fig. 10.8 Cross-section of the neck below the level of the sixth cervical vertebra. The section does not include the back half of the vertebra and associated spinal muscles (After *Gray*)

5 The compartments of deep fascia on each side, the *carotid sheaths* in which run the *carotid artery, internal jugular vein* and the *vagus nerve*.

Longitudinal section indicates the vertical arrangement, and the transition from larynx to trachea and pharynx to oesophagus (Fig. 10.10).

Triangles
The neck is divided into several triangles, by various muscles. The key is the *sternomastoid*, a long, flat muscle on each side, whose origin is by two heads from the inner end of the clavicle and upper border of the manubrium; it passes up and out to its insertion at the mastoid. The area in front of the muscle is the anterior triangle, and that behind the posterior triangle (Fig. 10.9).

The *anterior triangle* is bounded by the anterior midline of the neck and the anterior border of the sternomastoid, meeting at the apex below; its base, above, is the mandible. It is subdivided into

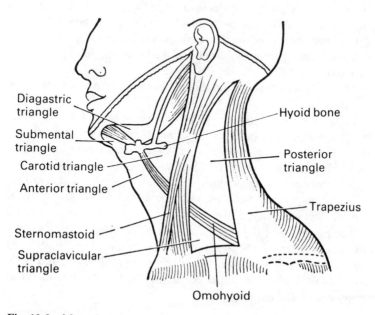

Fig. 10.9 Muscular triangles of the neck (After *Gray*)

four smaller triangles. The *muscular triangle* contains several small muscles which depress the larynx in swallowing; the *carotid triangle* contains the upper part of the common carotid artery as it divides into internal and external carotids, the internal jugular vein and important nerves; the *digastric triangle* contains the submandibular salivary gland; the *submental triangle* contains a few small lymph glands.

The *posterior triangle* is bounded by the back of the sternomastoid and the front of the trapezius, meeting at the apex at the occipital bone; its base is the clavicle. It is divided by the tiny omohyoid muscle into an upper *occipital* and a lower *supraclavicular triangle*. The former is mainly muscular, but the latter contains the subclavian vessels and brachial plexus of nerves, on their way to the arm.

Surface anatomy of the neck
Anterior. The *sternomastoid* is very prominent, especially when rendered taut by turning the head to the opposite side, when its sternal tendon can be felt between finger and thumb. Between the sternal heads of the two muscles is the *suprasternal notch*, a hollow at the upper border of the manubrium. The *external jugular vein* can be seen running obliquely down and back, superficial to the sternomastoid.

The main structures in the midline of the front of the neck are as follows.

In the hollow beneath the chin is the small *hyoid bone* felt when the throat is grasped between finger and thumb. Immediately below this is the *laryngeal prominence*, or Adam's apple, composed of the two halves of the thyroid cartilage meeting at a V-angle in front, with a notch at their upper border. Below this is the *cricoid cartilage* of the larynx, at the level of the sixth cervical vertebra; lower still are the cartilaginous rings of the *trachea*.

The pulsation of the *carotid arteries* is felt by pressure along the anterior border of the sternomastoids; behind this muscle, the posterior triangle is marked by a depression of the skin, and at the base of this, behind the upper border of the clavicle, the *subclavian artery* can be felt beating.

Posterior. At the back of the neck, the median longitudinal furrow between the muscle bellies overlies the spinous processes of the cervical vertebrae.

Fasciae and muscles of the neck
The deep fascia in the neck forms a continuous encircling sheet. There are also deeper concentric layers encircling the trachea and, deepest of all, the vertebral bodies. The muscles are partitioned by fascia, and a condensation of this membrane on each side forms the carotid sheath containing the common carotid artery, internal jugular vein and vagus nerve.

The cervical muscles are grouped as follows:

1 *Superficial: platysma, trapezius, sternomastoid.* The trapezius and sternomastoid are both supplied by the eleventh (*accessory*) cranial nerve and have already been dealt with. Each sternomastoid acts by bringing the ear down to the shoulder and turning the chin to the other side; both acting together flex the cervical spine, bringing the chin down to the sternum.

The *platysma* is a muscle of facial expression, a broad sheet without bony attachment. It arises from below the clavicle, and spreads up the side of the neck and over the mandible to blend with the facial muscles at the angle of the mouth.

2 The *suprahyoid* muscles are a small group between hyoid and mandible, forming the floor of the mouth and supporting the base of the tongue.

3 The *infrahyoid group* are strap-like muscles, descending from the hyoid bone and thyroid cartilage on each side of the midline to be attached to the back of the manubrium. They depress the larynx in swallowing.

4 The *prevertebral* muscles are the deepest, running in front of the vertebral bodies behind the other structures of the neck; they help flex the cervical spine.

5 The *scalene* muscles are found in the posterior triangle; they run down and out from the transverse processes of the cervical vertebrae to the first and second ribs. They elevate these ribs in inspiration, and are close to the subclavian vessels and brachial plexus as these run out over the first rib.

Pharynx

In the lower part of the neck, and in the chest, the air passage (larynx and trachea) and the food passage (oesophagus) are separate. But in the upper part of the neck, there is a common cavity extending as far as the base of the skull. This is the *pharynx*, situated immediately in front of the vertebral column.

In longitudinal section (Fig. 10.10) it can be seen that the pharynx lies successively behind the nasal cavity, the back of the mouth and the opening of the larynx, and it is therefore divided into *nasopharynx, oral pharynx* and *laryngopharynx*. Its upper end reaches the base of the skull, and it is continuous below with the oesophagus at the level of the sixth cervical vertebra.

The pharyngeal wall is a muscular cylinder designed to propel food between mouth and oesophagus, and is composed of three *constrictor muscles – superior, middle* and *inferior* – corresponding to the subdivisions of the cavity. In swallowing, it is essential to prevent food passing up into the nose round the back edge of the soft palate, or down into the larynx. Therefore, the soft palate is

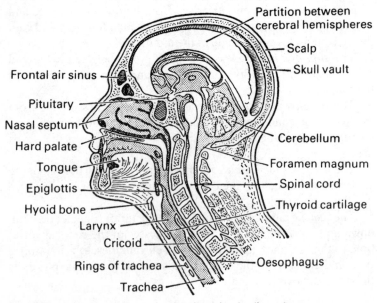

Fig. 10.10   Head and neck in longitudinal (sagittal) section

opposed to the posterior pharyngeal wall to shut off the nasopharynx, and the *epiglottis*, a flap of elastic fibrocartilage situated at the summit of the larynx behind the root of the tongue, is bent over backwards close the entry into the larynx. During the act, the larynx moves upwards and then recedes again (Fig. 13.1).

Larynx

The larynx is the organ of voice and part of the air passages. It opens above into the pharynx at the back of the tongue and is continuous with the trachea below. It projects forward in the upper part of the neck under the skin.

The organ is a framework of cartilages connected by ligaments and membranes to form a resonating chamber containing the vocal cords. The *thyroid cartilage* is the largest; its two leaves meet in the midline as the Adam's apple, with a superior notch. From the back of this notch arises the stalk of the *epiglottis*, a leaf-like fibrocartilage plate sticking up behind the hyoid and the base of the tongue. The *cricoid cartilage* at the bottom is a signet-ring-shaped structure, and immediately below is the first ring of the trachea.

The *vocal cords* are a pair of mucosal folds, strengthened by an underlying ligament, one on the deep surface of each thyroid leaf, stretching from front to back. The muscles associated with the larynx are *extrinsic*, e.g. the infrahyoid group, designed to move the organ as a whole as in swallowing, or *intrinsic*, concerned with regulation of tension in the vocal cords. The latter may lie relaxed at the sides of the cavity, meet in the midline and shut off the airway, or occupy an intermediate position, giving gradations of pitch. The interval between them is the *glottis*, the narrowest point of the respiratory tract.

The *trachea* or windpipe is a hollow cartilaginous and membranous tube, some 10 cm long, extending from the larynx through the thoracic inlet to its bifurcation into the bronchi. It is a ⌒-shaped tube, with the convexity under the skin of the front of the neck and the base resting on the oesophagus, which separates it from the vertebral column. It consists of a number of cartilaginous rings, incomplete behind, with connecting membranes. In the neck, the second, third and fourth rings are crossed by the isthmus connecting the two lobes of the thyroid gland. The common carotid arteries lie beside the trachea.

The trachea has a lining mucous membrane which is *ciliated* – i.e. each cell bears a hair-like process that wafts foreign irritants upwards toward the base of the tongue, where they are swallowed.

The *oesophagus* is a muscular channel about 25 cm long, for the passage of food between pharynx and stomach. It begins at the level of the cricoid cartilage as the continuation of the lower portion of the pharynx. It descends in front of the vertebral column behind the trachea and enters the thoracic inlet.

The *thyroid gland* is one of the group of endocrine glands. It lies in the lower part of the neck as a right and left lobe on each side of the trachea, with a connecting isthmus. Each lobe is pyramidal, apex upwards, and is applied to the sides of the cricoid and thyroid cartilages; laterally, it overlaps the carotid sheath. The gland is composed of numerous tiny vesicles containing *thyroxin*, the secretion essential for growth and activity; deficiency leads to cretinism in infants, and to retardation in adults. Any enlargement of the organ is called a *goitre*.

The *parathyroid glands* are four pea-like bodies embedded in the lobes of the thyroid, an upper and a lower on each side. They are ductless glands controlling the utilisation of calcium.

Great vessels of the neck
On the right side, the innominate branch of the aorta divides at the level of the sternoclavicular joint into subclavian and common carotid arteries; on the left, there is no innominate vessel, the subclavian and carotid arising directly from the aorta.

In the neck the *common carotid* ascends on either side of the trachea and thyroid gland to the upper border of the thyroid cartilage, where it divides into *internal* and *external* branches. It runs in the carotid sheath, with the internal jugular vein lateral to it and with the tenth cranial (*vagus*) nerve. Just below the bifurcation is a dilatation, the *carotid sinus*, where the vessel walls are very sensitive to changes in blood pressure; this sinus maintains a reflex control of the circulation to the brain. Apart from its terminal division, the common carotid has no branches.

The *internal carotid* continues to the base of the skull, enters the cranium and supplies the brain and the contents of the orbit.

The *external carotid* supplies the neck, and the outside of face and head. It gives large branches to thyroid gland, larynx and tongue and the facial artery, and ends by running up behind the ramus of the mandible in the substance of the parotid gland to finish as the superficial temporal artery to the scalp. It also supplies the main artery to the jaws.

The *internal jugular vein* corresponds to the common and internal portions of the carotid artery. It leaves the base of the skull as a continuation of the venous sinuses of the brain through a special foramen – in company with the ninth, tenth, eleventh and twelfth cranial nerves – and runs down the neck, lateral to the carotid artery and under cover of the sternomastoid. It ends by joining with the subclavian vein behind the sternoclavicular joint to form the innominate vein. There is no vein corresponding to the external carotid artery, but some veins from the face and scalp form the *external jugular vein*, which runs superficially over the sternomastoid and pierces the deep fascia above the clavicle to enter the subclavian vein.

The nerves of the neck

These fall into the following groups:

1 **Cranial nerves**. The ninth (*glossopharyngeal*), tenth (*vagus*), eleventh (*accessory*) and twelfth (*hypoglossal*) pairs of cranial nerves enter the neck from the cranium through foramina in the base of the skull. The first three travel with the internal jugular vein.

The *glossopharyngeal* is concerned with sensation and taste in the pharynx and back of the tongue, and pierces the wall of the pharynx to supply its mucous membrane.

The *vagus* runs down in the carotid sheath, between the vein and artery, giving branches to pharynx and larynx. Its role in the chest and abdomen and its place in the autonomic system are referred to later. An important branch to the intrinsic muscles of the larynx is known as the *recurrent laryngeal nerve*, for it hooks round the underside of the subclavian artery on the right to turn back to the larynx; the right recurrent laryngeal nerve is entirely within the neck, but the left descends to the thorax to hook round the aortic arch before ascending. Each recurrent nerve is intimately related to the back of the thyroid gland.

The *accessory nerve* supplies the sternomastoid and then runs down and back in the posterior triangle to reach the trapezius.

The *hypoglossal nerve* supplies the muscles of the tongue and the floor of the mouth.

2 **Cervical spinal nerves.** The roots of the spinal nerves emerge in pairs between the vertebrae on each side. The upper roots form a small *cervical plexus* supplying various muscles. The lower roots, C. 5, 6, 7, 8, together with the first thoracic root, form an important network, the *brachial plexus*, which lies in the lower part of the neck in the posterior triangle and gives rise to the nerves of the arm. This plexus forms a bundle with the subclavian artery and vein, that passes over the first rib to the arm. Thus the plexus is partly in the neck above the clavicle and partly in the axilla.

3 **Sympathetic chain.** In the neck, the sympathetic chain consists of three *ganglia* – the *superior, middle* and *inferior* – with a connecting *sympathetic trunk*; these lie deeply on the prevertebral muscles on each side, behind the carotid sheath. The trunk is continuous with the thoracic portion of the sympathetic chain.

Lymph glands of head and neck

The main collecting glands for the head and face are associated with the ear, the pre- and postauricular; those on the parotid and submandibular glands; the submental glands under the chin; the occipital glands at the skull base behind; and deep glands between pharynx and vertebral column.

There are superficial and deep chains of glands and lymphatic vessels. The superficial glands are grouped around the external jugular vein and accessory nerve in the posterior triangle; the deep glands surround the internal jugular vein and carotid.

At the base of the neck, on the right, a major lymphatic trunk is formed from the vessels draining the right arm and right side of head and face; this discharges into the union of subclavian and internal jugular veins. On the left side, the thoracic duct, with all the lymph from both legs, and from the abdominal and thoracic viscera, arrives in the posterior triangle, adds to itself the vessels of the left arm, head and neck, and empties into the corresponding venous angle. The thoracic duct is much larger and more important than the right lymph trunk, but even when distended is no thicker than a matchstick.

11

Regional Anatomy: the Spinal Column

The *spine* is the bony axis of the body and is composed of individual vertebrae in a segmental pattern. These articulate with each other, and the sum of the limited motion between individual pairs is considerable. The column is traversed by the central canal, which encloses the spinal cord; and it supports the weight of the trunk and transmits it to the legs.

The vertebrae

The vertebrae are grouped regionally as:

Cervical (neck)	7
Thoracic (chest)	12
Lumbar (abdominal)	5
Sacral (hip)	5
Coccygeal (tail)	4

Although modified in the different regions, the essential features of a vertebra are similar everywhere (Fig. 12.2). In front a massive, rounded *body* projects forward, with upper and lower surfaces facing the vertebrae above and below. Behind is the *neural arch*, which, with the back of the body, forms a complete bony ring – the *neural canal* for the spinal cord. The arch is attached to the body by two pillars or *pedicles*; projecting at either side are the two *transverse processes*; and the *spinous process* sticks out backwards and downwards from the back of the ring under the skin of the back. There is a pair of small *articular facets* on the upper surface of the

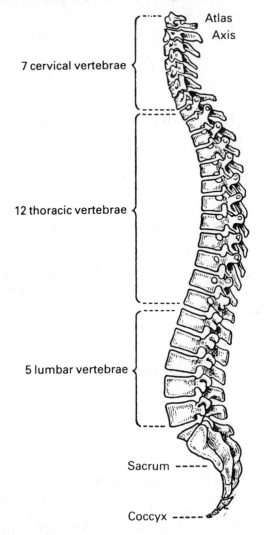

Fig. 11.1   The spinal column, lateral view

arch and a pair below for articulation with adjacent vertebrae. The vertebral body consists of spongy bone, with a thin shell.

A pair of vertebrae fit together to leave on each side an *intervertebral foramen* for the exit of a spinal nerve. There are thirty-four

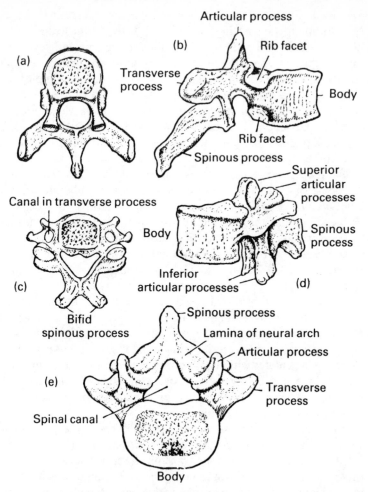

Fig. 11.2 Some typical vertebrae. (a) Thoracic vertebra. (b) Thoracic vertebra, side view. (c) Cervical vertebra. (d) Lumbar vertebra, side view. (e) Lumbar vertebra, from above

pairs of spinal nerves, since the first cervical nerve emerges between the base of the skull and the uppermost cervical vertebra, while there is only one coccygeal nerve below.

The *cervical* vertebrae are delicate, with small bodies, and a very large neural canal, as the cord is thickest at its upper end before it has given off most of its roots. The transverse processes are pierced by a foramen transmitting the *vertebral artery*, which runs up the neck from the subclavian and enters the foramen magnum to supply the hindbrain. The spinous process is short and bifid. The first cervical vertebra is the *atlas*, supporting the head. It is a simple bony ring without a body, and its upper surface articulates with the condyles of the occipital bone to form the joint at which nodding occurs. The second cervical vertebra is the *axis*; from its upper surface a peg-like process sticks up within the atlas ring. Skull and atlas rotate on this peg to turn of the head.

The *thoracic* vertebrae increase in size downwards. The bodies are heart-shaped, with facets at the sides for articulation with the heads of the ribs; a similar facet on the transverse process articulates with the neck of the rib. The neural canal is relatively small; the spinous process is long and projects downward.

The *lumbar* vertebrae are under considerable load and so have massive bodies. The neural canal is triangular and the squat, spinous processes project horizontally backwards.

The *sacral* vertebrae are fused into one bone, the sacrum, intervening between the innominate bones as the posterior segment of the pelvic ring. It is a large, flattened, triangular bone with anterior (pelvic) and posterior (subcutaneous) surfaces (Fig. 11.3). The base articulates with the fifth lumbar vertebra at the lumbosacral joint, the apex below with the coccyx. It is placed obliquely, with its long axis directed backwards, and the anterior surface is the hollow in which lies the rectum.

Although a single mass, the general features of the component vertebrae are still obvious, the bodies being indicated by transverse ridges. Four pairs of foramina transmit the sacral nerve roots, and the body of the bone is traversed by the sacral canal, which, at this level, contains not the spinal cord (this ends in the upper lumbar region) but a mass of nerve roots.

Above, the promontory of the sacrum is a prominent ridge in the anterior surface which forms the back of the pelvic brim. On each side a wing or *ala* reaches out to form with the innominate bone the *sacroiliac joint*, a large interlocking pair of surfaces with little motion. The back of the sacrum is the lower origin of the sacrospinalis muscle.

The *coccyx* is the rudimentary tail appendage, a small irregular

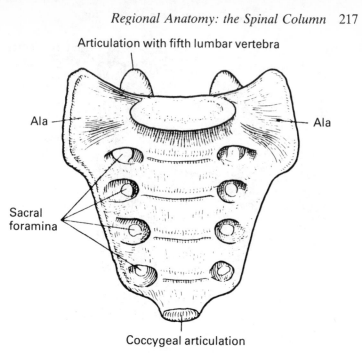

Articulation with fifth lumbar vertebra

Ala

Ala

Sacral
foramina

Coccygeal articulation

Fig. 11.3   The sacrum, pelvic aspect

mass of four fused vertebrae. It is triangular, angled forward at the
sacrococcygeal joint, and quite solid.

The spinal column as a whole

The vertebral bodies are separated by thick cushions of fibrocartilage, the *intervertebral discs*. These increase in size down the column
until in the lumbar region they are 13 mm thick. Together, they
make up a fourth or fifth of the length of the column. Each consists
of an outer fibrous ring enclosing a gelatinous core under tension,
the pulpy *nucleus*, which acts as a ball-bearing. The intervertebral
joint consists of the disc between the bodies, plus the simple plane
synovial joints between the pairs of articular processes on the
arches. Long strap-like *ligaments* bind the bodies, the anterior and
posterior ligaments which traverse the length of the column on the
front and back of the bodies; shorter bands connect the spinous and
transverse processes, to make a flexible entity.

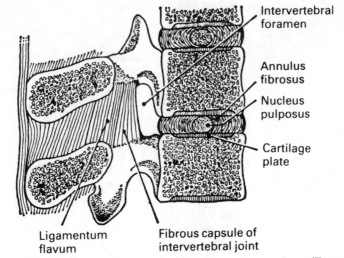

Fig. 11.4   Median section of articulating thoracic vertebrae (From *A Companion to Medical Studies, Vol. 1*)

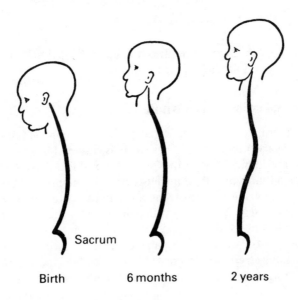

Fig. 11.5   Spinal curves at different ages

Although the spine is quite straight looked at from in front or behind, there are in side view a series of curvatures alternating in direction. The cervical and lumbar curves are forward convexities or *lordoses*; the thoracic and sacral curves are forward concavities or *kyphoses*. This arrangement of arcs, spanned by the spinal muscles, is sounder than a more rigid, straight column. The child's spine at birth does not possess the cervical and lumbar curves, only the anterior concavity of the whole spine (the crouched foetal position) and that of the sacrum. As the head is held up in the first six months, the cervical lordosis develops, and the lumbar curve results from sitting up and standing in the second year. In the adult, the vertebral bodies are somewhat wedge-shaped to conform to the shape of the curves.

Spinal movements

These are:

1 *Flexion*, forward bending, maximal in the cervical region, though considerable in the lumbar spine;
2 *Extension*, backward bending, maximal in the lumbar region but also taking place in the neck;
3 *Rotation*, a twisting of the column on the longitudinal axis traversing the nuclei of the discs, freest in the upper thoracic region and negligible elsewhere;
4 *Sideways bending*, maximal in lumbar and cervical regions;
5 *Circumduction*, a swaying movement combining all the above, in which the trunk describes the surface of a cone, apex down.

12

Foodstuffs and Vitamins

Food and energy

The energy of *light*, absorbed by plants for photosynthesis, initiates every biological process; the *heat* given off by animals and plants is the final waste. The conversion of one to the other permits useful work. The bound chemical energy in working muscles is transformed, partly into work, and partly into heat. The basic unit of heat is the calorie, the amount of heat required to raise the temperature of 1 gramme of water from 14·5°C to 15·5°C, equivalent to $4·2 \times 10^7$ ergs. (Under the SI system, 1 kcal (1000 cal) = 4·186 megajoules.)

The fuels of the body are protein, fat and carbohydrate. Most of the energy in their carbon and hydrogen linkages can be made available by oxidative metabolic processes. The heats of combustion of these foods measured in the laboratory are slightly larger than the energy actually available to the body, since digestion and absorption are incomplete and oxidation is only partial. Nevertheless, 99% of the carbohydrate, 95% of the fat, and 92% of the protein taken are absorbed; only 5% of the calorie intake is lost in the faeces. Since much of the nitrogen intake is excreted in the urine as urea, uric acid and other substances, about a quarter of the calorie value of ingested protein is lost. The amounts of heat available from 1 gramme of a basic foodstuff are:

1 g of protein	4 kcal
1 g of fat	9 kcal
1 g of carbohydrate	4 kcal

It is possible to calculate the energy value of a diet by analysing its components and applying these factors (Table 12.1).

Table 12.1 Energy value and nutrient content of common foods (per gramme)

	Water %	Protein %	Fat %	Carbohydrate %	Energy kcal
Wheat flour	15·0	13·6	2·5	69·1	3·4
White bread	38·3	7·8	1·4	52·7	2·4
Rice	11·7	6·2	1·0	86·8	3·6
Milk	87·0	3·4	3·7	4·8	0·7
Butter	13·9	0·4	85·1	trace	7·9
Cheese	37·0	25·4	34·5	trace	4·3
Steak	57·0	20·4	20·4	nil	2·7
Haddock	65·1	20·4	8·3	3·6	1·8
Potatoes, raw	80·0	2·5	trace	15·9	0·7
Peas	72·7	5·9	trace	16·5	0·9
Cabbage, boiled	96·0	1·3	trace	1·1	0·1
Orange	64·8	0·6	trace	6·4	0·3
Apple	84·1	0·3	trace	12·2	0·5
Sugar	trace	trace	trace	100·0	3·9
Beer	96·7	0·2	trace	2·2	0·3
Spirits	63·5	trace	nil	trace	2·2

These foods not only supply energy, but also replace wear and tear in the tissues. Whereas any foodstuff can be used for production of energy, only proteins can be used for cell building. For energy production, fat is more economical than the others, since it provides twice as much heat and it is used in much purer form.

Proteins are not an economical way of obtaining energy; fats and carbohydrates are normally used to meet energy requirements and proteins to repair tissue breakdown. A man would have to eat 2 kg of lean meat daily to obtain his basic minimum of 2000 calories, an impossible amount to digest. Nor is protein stored to any extent, except during growth. Protein, though an essential of the diet, has no advantages as a source of energy over fats and carbohydrates, which are capable of considerable interchangeability.

Nevertheless, it is an advantage to include as much protein in the diet as possible; surplus calories from fat and carbohydrate are sometimes referred to as 'empty' calories.

The diet

The average daily energy requirement of a working man is 3000 calories. An adult woman needs 2500 calories, because of her smaller size and surface; and sedentary workers need less. Because of loss of calories in preparing food, and in digestion, an extra allowance is necessary and the calorific value for an adult man, *as purchased*, is some 3500. To maintain health a diet should also contain protein, fat and carbohydrate in the proportions indicated below; enough fresh food to supply the necessary vitamins; enough mineral content; and *it must be palatable.*

Protein should supply one-sixth of calorie needs, i.e. at least 100 g a day, at least half in the form of first-class animal protein. An excess serves no particular purpose, even for heavy workers. What protein does, more than other foods, is to give a fillip to metabolism generally, which increases heat production and keeps one warm in cold weather.

Carbohydrates form the bulk of most diets; they are cheap and easily obtainable as bread and other filler foods. The optimum daily need is some 500 g. There is an absolute minimum; below this the carbohydrate:fat ratio is disturbed, fats are not completely burnt, and acid products accumulate in the tissues.

Fat is valuable because of its high energy value. A normal diet contains only one-seventh of its weight of fat (some 100 g), but this supplies a quarter of the total calories; its concentration is exaggerated by the small amount of water it contains compared with other foods. Although interchangeable with carbohydrate as an energy source, fat has certain advantages. Its slow absorption prevents the hunger which would be felt with a largely carbohydrate diet; and for really heavy work in cold or Arctic conditions a large amount of fat is the only practical way of supplying calorie needs.

Respiratory quotient

We can deduce which substance is being oxidised in the body to provide energy by comparing the oxygen used up (contained in the inspired air) with the carbon dioxide evolved (got rid of in the expired air). Thus the burning of carbohydrates may be represented by the equation:

$$C_6H_{12}O_6 + 6O_2 = 6CO_2 + 6H_2O$$

where the respiratory quotient $\dfrac{CO_2 \text{ expired}}{O_2 \text{ inspired}}$ is exactly 1.

Since different materials are burnt at the same time, the actual respiratory quotient depends on the relative proportions of the foodstuffs, whose respective quotients are:

Carbohydrates   1·0
Protein   0·8
Fat   0·7

The actual quotient is always less than 1, since a purely carbohydrate diet never exists; the quotient represents the gross result of oxidation of the average mixture of foodstuffs.

For energy any of the foodstuffs may be utilised, and carbohydrates and proteins are called *isodynamic* because an equivalent weight of each produces the same amount of heat. Carbohydrates and fats in the proportion of rather more than 2:1 also provide the same number of calories, and can replace each other in these proportions without altering the available energy. Although protein is theoretically interchangeable in the same way, an upper limit is set by the capacity for its digestion. Since protein is not stored, any excess is broken down and excreted as nitrogenous compounds in the urine and faeces, i.e. there is a normal state of nitrogenous equilibrium. It is different with fats and carbohydrates, for an excess of either is stored in the body as *fat*.

Protein is broken down by digestion into its constituent amino-acids; these enter the bloodstream, where they partake in equilibrium with the protein of the body cells, the latter taking up such as they require for repair. Residual amino-acids are disintegrated in the liver; their nitrogenous part is lost as urea through the kidneys, the rest burnt to carbon dioxide and water. Only some proteins contain those essential amino-acids which cannot be synthesised in the body; these are the *first-class proteins* of meat, wheat, and so on. The *second-class proteins* lack some of the essential amino-acids, or are made up of inessential ones; for example, gelatin. A deficiency of essential acids results in loss of appetite and failure of growth; to make this provision animal protein is more economical than vegetable, though the latter may be adequate if the diet is varied.

Foodstuffs in detail

Foods may be grouped as follows: cereals; starchy roots; pulses and legumes; vegetables and fruits; sugars and derivatives; meat, fish

and eggs; fats and oils; and beverages. No one is essential, but a
good diet contains foods of all classes.

Cereals
These form the chief food of most human beings; even in rich
countries, they provide a quarter of the total calories and are the
largest item of diet. All cereal seeds have much the same structure
(see Fig. 12.1). The main bulk is the starchy endosperm, with the
germ at one end. Outside are the outer endosperm, aleurone layer
and pericarp, enclosed in a woody husk. The grain contains 80%
starch, a little fat and as much as 10% protein; but the protein of any
one cereal is poorly suited to human requirements, so a mixture of
sources is required. Cereals contain much calcium and iron, but the
phytic acid present forms insoluble phytates. *Whole* grains are a
valuable source of B vitamins. They contain no vitamin C, A, or D;

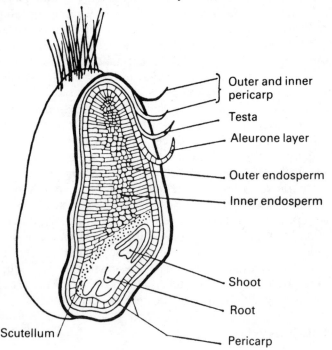

Fig. 12.1   The structure of the wheat grain (From McCance, *Lancet*
i (1946), 77)

but oily grains like maize contain vitamin A precursor and much vitamin E.

Wheat and rice are usually eaten after milling to white flour or rice. The material removed contains most of the B vitamins and thiamine, and therefore rice-eaters may suffer from beriberi (p. 230). The losses can be made good by enriching white flour; it is difficult to enrich polished rice.

The poor man's cereal is *millet*, which is not highly extracted. *Maize* is an excellent food, although if it forms the bulk of the diet it may be associated with pellagra (p. 230). *Wheat* is the only cereal (apart from rye) that can be made into bread.

The nutritive value of cereals is considerable. Bread and rice provide calories, proteins, B vitamins, calcium and iron. Only when a diet consists of an excess of cereals, or of any one cereal in particular, does it become unbalanced.

Starchy roots
The potato is an excellent, energy-rich food. On no other single food can a man survive indefinitely, for potatoes contain significant quantities of vitamin C and carotene, and adequate amounts of the B vitamins. Tropical roots are the cassava and yam, a rich source of energy, though their protein content is not high. Children fed largely on them are liable to protein-deficiency diseases.

Pulses and legumes
Pulses, lentils and dhals have a protein content double that of cereals, and are valuable in the tropics. They are also rich in B vitamins. The *soya bean* contains up to 20% of fat.

Vegetables and fruits
These include leaves, flowers, seeds, stems and roots. They contain much water, few calories and little protein; but they are filling and prevent constipation. They contain some vitamin B and C, but the latter is often lost in cooking. The chief vitamin is carotene, the precursor of vitamin A. Vegetables increase the nutritive value of cereal diets in the tropics but contribute little to the average Western diet.

The main nutritional value of *fruits* lies in their vitamin C content. The banana provides carbohydrate in abundance but lacks protein.

Sugar

There has been a great increase in sugar consumption and this may be related to the increased incidence of cardiovascular disease. Without other nutrients, sugar provides 'empty' calories, and promotes dental caries, obesity and diabetes.

Meat, fish and eggs

Meat, with its fat, contains one-fifth fat, three-fifths water, and one-fifth protein of high biological value. It is also a source of iron and the B vitamins, but not of vitamins A and C. *Offal*, such as liver and kidneys, has the same general value as meat but more vitamin A and iron. *Fish* is a rich source of protein, oil, and vitamins A and D. *Eggs* consist of 10% first-cass protein, 10% of fat, much vitamin A, and calcium and iron. They are a most useful item of diet, though their high cholesterol content is unpopular in some quarters.

Milk

Human milk is a complete infant food. Cow's milk contains more protein and less carbohydrate, and can be made suitable by diluting and adding sugar.

Table 12.2   Milk values (per 100 grammes)

	Human milk	*Cow's milk*
Calories	70	66
Carbohydrate	7 g	5 g
Protein	2 g	3·5 g
Fat	4 g	3·5 g
Calcium	25 mg	120 mg
Phosphorus	16 mg	95 mg
Iron	0·1 mg	0·1 mg
Vitamin A	170 units	150 units
Vitamin C	3·5 mg	2·0 mg
Vitamin D	1·0 units	1·5 units
Thiamine	17 μg	40 μg
Riboflavine	30 μg	150 μg
Nicotinic acid	170 μg	80 μg

Milk is rich in calcium, but poor in iron, and infants may develop anaemia after six months if mixed foods are not added. If milk is boiled, so destroying vitamin C, scurvy may develop. Mothers usually produce adequate milk of good quality, even at their own expense, as when the calcium content of the milk is derived from the maternal skeleton. Milk is probably essential for growing children, but not for adults. *Skimmed milk*, with the butter fat removed, can be made as a stable powder, and is a valuable protein source for children in sub-starvation areas, provided it is not made up with contaminated water. *Cheese* is made by clotting milk with rennet; it contains most of the original protein and some of the fat, and is highly nutritious.

Fats and oils

These may be animal or vegetable. The former are mainly dairy fats, beef suet and pork lard; the latter include oils of the olive, groundnut, cottonseed and red palm, and corn oil. Margarine is made by mixing vegetable oil with whale oil. Animal fats (except those of fish) are solid and rich in saturated fatty acids. Vegetable oils are liquid and contain a lot of unsaturated fatty acids, and so does fish oil. Fats are a concentrated source of calories, keep the diet from being too bulky, and are vehicles for the fat-soluble vitamins. They are valuable in cooking, and in rich countries far more is taken than is required to supply the essential fatty acids. Butter contains vitamins A and D, which have to be added to margarine; but there is little to choose between the two in nutritive value. Vegetable oils contain no vitamins.

Beverages

These are drunk for their flavour or alcohol content. *Ethyl alcohol* is produced by fermenting grain or fruit. It is a nutrient, readily absorbed, and an immediate source of energy. Its surplus calories may cause obesity. It is a sedative, and promotes well-being in moderation, but may become addictive and have toxic effects on the nervous system if taken in excess.

Beer is made by converting grain starch to maltose by maltase, and then fermenting it by yeast. Beer contains up to 6% alcohol, but no vitamins, except riboflavine. *Wine* is made by fermenting grapes or other fruits; it contains 8%–10% of alcohol and may be fortified with added alcohol up to 15%. It may contain large amounts of iron, causing chronic liver disease. *Spirits* are made by distilling fermented liquor; they may contain some 30% of alcohol but are devoid of other nutrients. Imperfect distillation may yield spirits containing methyl alcohol, which is very poisonous and can cause blindness or death.

Coffee and *tea* contain caffeine, theophylline and theobromine. Caffeine stimulates the nervous system and increases the output of urine. Both contain tannins, which are protein precipitants, and may cause gastritis. Although tea and coffee are of no nutritive value, they are usually taken with milk and sugar and this may be a hidden source of calories.

Soft drinks may contain vitamin C; their calorie value depends on added sugar. *Fruit juices* are rich in potassium and useful to convalescents. *Mineral waters* have a varied but insignificant mineral content.

The vitamins

If an animal is fed on a diet with adequate amounts of fat, carbohydrate, protein and salts, *but in which these constituents have been purified*, it becomes ill, stunted and dies. This may be entirely prevented by the daily addition of a teaspoonful of milk. This is because natural foods contain minute quantities of organic substances known as *vitamins* that the body is unable to synthesize (except, to some extent, vitamin D). Many are components of essential enzyme systems. They are essential for the proper utilisation of food, normal growth and to maintain health in the adult. Yet they contribute nothing towards energy requirements or tissue repair. There are five major diseases due to lack of one or other vitamin: scurvy, beriberi, pellagra, keratomalacia and rickets, still major scourges in the third world.

The vitamins are either fat-soluble or water-soluble. The water-soluble vitamins cannot be stored for they are readily excreted by the kidneys, and deficiency disease soon arises if the intake is inadequate. The fat-soluble vitamins are stored in adipose tissue, particularly the liver, and a well-fed individual will remain healthy for many weeks even if his external supply is cut off.

Fat-soluble vitamins
Vitamin A (retinol). This is present only in certain animal foods – dairy produce, liver, and particularly fish-liver oils, although it was originally formed in green plants, either in pasture or the diatoms of

the sea. The small intestine can also manufacture it from the carotene pigment in some plant sources. In the UK half the vitamin A is obtained from animal sources, and half from precursors in plants, and deficiency disease is rare. But it is common in the tropics, where it causes stunting, susceptibility to disease, defective repair of epidermal tissues and progressive collapse and disorganisation of the eye – all from lack of, say, 1 mg of vitamin A a day. Pure vitamin A is pale yellow, insoluble in water, a highly unsaturated aromatic hydrocarbon stable in ordinary cooking.

Vitamin D (calciferol). This, too, is found in dairy produce, and in eggs and liver. It is also formed in the skin from the action of the ultraviolet rays of sunlight on a sterol compound, so the body is not entirely dependent on dietetic sources. However, heavily pigmented children from the tropics can develop rickets when moved to a smoky temperate zone.

Vitamin D promotes absorption of calcium from the bowel and deposition of bone by the formation of calcium salts. It has a major influence on the growth and hardening of bones and teeth, and any deficiency in childhood leads to rickets and dental caries. Rickets is a disease in which soft, poorly calcified tissue is laid down instead of normal bone so that deformities develop – spinal curvature, bowlegs or knock-knees. It is readily cured by moderate intake of the vitamin. There is an adult form of rickets called *osteomalacia* – usually occurring in elderly women living on tea and bread and butter whose vitamin D intake is low – and again, deformity and fractures of the bones are common and curable. The vitamin can be made artificially by irradiating ergosterol, and is toxic in large doses.

Vitamin E. This is essential to normal fertility. There is no evidence that vitamin E deficiency ever occurs in humans, but in rats fed on artificial diets the females may abort, the males become sterile, and embryos fail to develop. The vitamin is found in quantity in oil of wheat germ.

Vitamin K. This is a complex of naphthoquinone derivatives essential to the complex mechanism of blood clotting. It can be synthesized. Dietary deficiency does not exist.

Water-soluble vitamins

Vitamin B. This is a complex of three factors:

1   *Vitamin B₁ (thiamine)* is abundant in the husks of cereal grains, pulses, yeast, milk and meat; it is lost when wheat and rice are subjected to machine milling. Thiamine catalyses the oxidation of carbohydrates in cells, particularly in nerve cells, and a deficiency causes beriberi. In *dry beriberi* there is peripheral neuritis, muscle wasting and paralysis; in *wet beriberi* there is heart-failure and waterlogging of the tissues. Wherever rice is a staple and finely milled, there is a risk of the disease, and even in Western countries synthetic thiamine is often added to white flour. The required daily intake is about 1·5 mg.

2   *Nicotinic acid* is another member of the B complex. Deficiency causes *pellagra* – dermatitis, dementia and diarrhoea – associated with the exclusive consumption of maize which is a poor source of nicotinic acid. But, as it happens, the body can synthesize it from the tryptophane in rice, so that rice-eaters are protected even though rice itself contains little of the natural factor. Pellagra may also result from interference with intestinal absorption, in chronic alcoholism. Nicotinic acid is present in whole cereals and rice, in meat, offal and fish; it is abundant in yeast and yeast extracts but not at all in dairy products, fruit or vegetables. In the UK and the USA, it is legally required to add it to white flour. The mean necessary daily intake is of the order of 10–12 mg.

3   *Riboflavine* is a yellow pigment, an essential co-enzyme for tissue respiratory processes. It is present in most foods. Lack of riboflavine may cause ulceration at the corners of the mouth. The daily dietary requirement is 1–2 mg.

The anti-anaemic B vitamins. Red blood cells only live for 100 days or so and must be continually reproduced. Certain vitamins are essential to this process, and if they are inadequate there will be anaemia.

Vitamin B₁₂ (cyanocobalamin) occurs in animal foods but not plants. It is a complex porphyrin derivative containing cobalt. It is a by-product in the manufacture of certain fungal antibiotics. The daily requirement is less than 1 μg. For proper absorption by the intestine, an *intrinsic factor* secreted by certain stomach glands is essential; when this is absent, the individual develops pernicious anaemia. This may be cured by eating large amounts of animal liver, or, more usually, by regular injections of the vitamin, on which the patient is as dependent as a diabetic is on insulin injections.

The other anti-anaemic B vitamin is *folic acid*, whose main source is the leaves of green vegetables. The average daily requirement is some 50 μg. Anaemia due to deficiency is common in tropical countries, particularly in pregnancy.

Other water-soluble vitamins, such as pyridoxine and pantothenic acid, are components of essential, tissue-enzyme systems. They are widely distributed and deficiency is rare. But the most important remaining water-soluble factor is ascorbic acid – *Vitamin C*.

Vitamin C. Ascorbic acid is a crystalline sugar, readily synthesized, soluble and easily oxidised. The main source is from fruit, vegetables and milk. Meat and eggs contain only a trace; potatoes are a useful source. It is easily destroyed in cooking, and when diets contain little fruit or vegetables, or vegetables are boiled, scurvy may result, characterised by haemorrhages such as bleeding gums, and also by indolent healing of wounds and fractures. Scurvy can be readily prevented with citrus fruits. Milk contains little ascorbic acid and what exists is destroyed by boiling, so infants are particularly liable to the disease. The minimum daily intake is about 25 mg, provided by an ordinary helping of cabbage or other green vegetable, a good helping of potatoes twice daily, or by 40 g of fresh oranges or orange juice.

To sum up, most vitamins are of known composition and can be synthesized or obtained in pure form. The only evidence of their effects is that when the body is deprived of them, deficiency disorders appear. Although essential to health, they are not foodstuffs and will not compensate for an inadequate diet. Further, in adults living on a good mixed diet, there is no need for the addition of vitamin preparations. This is not the case with growing children, where supplements of fish oil and orange juice may be desirable. The chronic alcoholic, because his drug interferes with the absorption and storage of vitamins, may need large doses of multi-vitamin preparations.

13

Digestion

The alimentary tract is a long, muscular tube, extending from mouth to anus, lined with an epithelium which secretes digestive juices and absorbs the products of digestion. This lining is the mucous membrane. There are also major glands lying outside the bowel altogether – the salivary glands, liver and pancreas – whose secretions are discharged into the intestine. These secretions amount to many litres in twenty-four hours and most of this fluid is reabsorbed; so vomiting or diarrhoea may cause profound dehydration and electrolyte disturbance. Secretion takes place mainly in the upper part of the tract, absorption in the middle zone and excretion in the lower segment.

Digestion and absorption are unconscious processes; bowel movement is involuntary and effected by smooth muscle; but chewing, the initiation of swallowing, and defaecation are conscious processes.

The *object of digestion* is to convert complex and insoluble constituents of food into simpler substances that can dissolve and diffuse through the lining of the intestine to enter the blood or lymph. Thus carbohydrates must be reduced to the simplest monosaccharide form or the cells will not be able to assimilate them. Fats must be hydrolysed to fatty acids and glycerol – though some enter the lymph unhydrolysed as finely emulsified globules – and proteins are split into their constituent amino-acids. In dealing with these, the body displays considerable ingenuity; to convert the protein of cheese into a human form of protein is like taking a meccano model to pieces and picking out the right bits to build a new one. Digestion is accomplished by the enzymes in the juices formed in the glands along the alimentary tract. Each foodstuff needs a specific enzyme to break it down.

The mouth

Entry of food into the mouth promotes a reflex flow of saliva. The stimulation of the taste nerve-endings is transmitted to the brain and relayed along the nerves controlling salivary secretion. But there may also be a 'psychic' secretion on the mere anticipation of food, and this is a 'conditioned' reflex in which an association of ideas has the same effect as the physical stimulus.

The saliva is a mixture of the secretions of three pairs of salivary glands – the *parotid* in the cheek, the *submandibular* below the angle of the jaw and the *sublingual* beneath the tongue – as well as of many minute glands in the mucous membrane of the mouth and palate. It contains *mucin*, which makes the food easy to swallow; *ptyalin*, an enzyme that digests starch to maltose; and minerals, particularly *calcium phosphate*, which is deposited on the teeth as tartar. Much of salivary digestion actually occurs in the stomach. Food does not remain long in the mouth, but the swallowed mass still contains ptyalin, which continues to act until penetration of the bolus by the acid gastric juice brings the process to a halt. There is normally a continuous slight secretion of saliva which is suppressed by fear and by drugs that dry the mouth. No absorption of nutrients occurs in the mouth.

Chewing breaks up the food and mixes it with saliva, and small amounts are passed at intervals into the pharynx to be swallowed. Swallowing is initially a voluntary, but eventually an involuntary, activity.

Swallowing

The oesophagus is a long muscular tube, beginning behind the lower part of the larynx and travelling down the back of the thoracic cavity to pierce the diaphragm and join the stomach. Chewing produces a round *bolus* of food which is passed down to the stomach in the act of swallowing. In swallowing it is essential to prevent food passing upwards into the nose round the back of the soft palate, or down

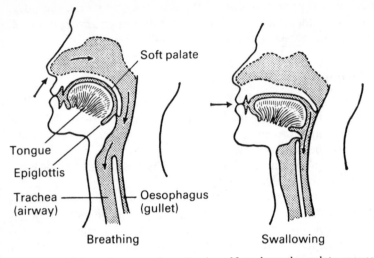

Tongue
Epiglottis
Trachea (airway)
Soft palate
Oesophagus (gullet)

Breathing   Swallowing

Fig. 13.1 The mechanism of swallowing. Note how the palate moves upward to block the back of the nose and the epiglottis down to shut off the windpipe

into the lungs. Therefore the soft palate is apposed to the back wall of the pharynx to seal off the nasal cavities, and the epiglottis (the elastic flap of fibro-cartilage situated behind the root of the tongue) bends backwards to close off the larynx. At the same time, the larynx as a whole moves upwards and then descends again (Fig. 13.1). The initiation of swallowing is largely voluntary, but once the food touches the pharynx, soft palate and epiglottis, the process continues as a reflex. In the oesophagus, the food is propelled by *peristalsis*, a wave of contraction controlled by the nerve plexuses in the oesophageal wall. At the lower end, the cardiac sphincter relaxes reflexly and food enters the stomach. This sphincter is normally closed to prevent reflux of acid gastric juice. The mean transit time along the oesophagus is some two seconds.

The stomach

The stomach is a hollow, muscular organ lined with a glandular mucous membrane which secretes the gastric juice. It acts as a reservoir in which a mass known as the *chyme* is formed from the ingested food; digestion continues until the chyme is transferred to the duodenum, but there is little absorption, except for water, alcohol and simple crystalloids.

Small amounts of gastric juice are present in the resting state. The anticipation of food, or its entry into the mouth, excites a reflex or conditioned secretion. More is formed when food actually enters the stomach; a pint of highly acid juice may be secreted after a heavy meal, the process continuing up to four hours. The stimulus to secretion is not mere contact with food but a hormone, *gastrin*, which is released into the circulation from the pyloric region of the stomach, and stimulates the body of the organ. Protein, alcohol and coffee are potent excitants of this phase of secretion. The secretion is strongly acid, with up to 0·5% of hydrochloric acid; some of which is neutralised by mucus, so only some 0·3% is present in the free state.

The enzymes present are *pepsin*, which converts proteins into polypeptides, and *rennin*, which curdles milk into a casein clot, to be digested by pepsin. Gastric juice acts by virtue of its acidity and through its enzyme action. The acid alone is responsible for minor breakdown of certain foods, but most of the work is done by the pepsin, which also helps liberate the fat of the food by dissolving the fibrous framework round its globules, preparing it for digestion in the duodenum.

The empty stomach is not entirely still; small contractions travel from the cardiac end towards the pylorus. A meal distends the organ and this stimulus excites a series of contraction waves which propel the food from the fundus towards the pylorus on the right. For the first thirty to sixty minutes after a meal, the opening into the duodenum at the pyloric sphincter remains closed; thereafter, it opens at intervals and permits gastric contents to pass in intermittent spurts into the duodenum. By the end of gastric digestion, in three to four hours, all the food passed through; the sphincter relaxes, and some duodenal contents and bile regurgitate into the stomach. Fluids leave the stomach long before solids and may enter the duodenum in a few minutes; fats remain much longer than carbohydrates.

The stomach is not perhaps a very important organ in digestion and absorption, as digestion can take place efficiently without it. Its chief role appears to be to receive and store food at irregular intervals, and to pass on more regularly partly digested food in smaller quantities to the small intestine, where most digestion takes place.

The mobility of the muscular wall and the rate of secretion are so intimately affected by emotional states that it is not surprising that dyspepsia or ulceration are often of nervous origin.

Small intestine

It is here that most of the digestive process occurs, effected mainly by the secretions of the liver and pancreas, which discharge into the duodenum. The principal function of the bowel wall is absorption. The three digestive juices, are the *pancreatic juice*, the *bile* formed by the liver and the intestinal juice produced by the lining of the bowel. The bile and pancreatic juice are discharged into the duodenum via ducts from the pancreas and liver, which open in common; the intestinal juice is formed along the entire length of the small intestine, but plays little part in the digestive process. It does contribute water to aid the solution and absorption of digested food. Secretion and digestion predominate in the upper coils of the small intestine and absorption of digested products into the portal circulation in the lower; but both processes go on simultaneously everywhere.

Pancreas

This organ secretes the pancreatic juice into the duodenum; but there are small islets of contained tissue that produce the hormone *insulin* (see p. 254).

The pancreatic juice is formed in two ways. There is a response to the ingestion of food due to a nervous reflex; but there is a more important response to the entry of acid gastric contents into the small intestine. This is a beautiful example of the work of chemical messengers, or *hormones*, in the body. When stomach contents enter the intestine, some of the acid extracts from the bowel wall a potent stimulating agent called *secretin*, which enters the bloodstream, is carried to the pancreas, and stimulates the secretion of pancreatic juice. Alkaline pancreatic juice then enters the duodenum and neutralises the acid influx, thus automatically bringing to a stop the mechanism for evoking pancreatic secretion. The next spurt of gastric contents starts a fresh cycle, and so on. Thus pancreatic secretion is exactly suited to the needs of digestion.

The pancreatic juice is as alkaline as gastric secretion is acid. It contains three enzymes: *trypsin*, which furthers the digestion of proteins to amino-acids; *amylase*, which hydrolyses carbohydrates to glucose or disaccharides; and *lipase*, which breaks up fats into fatty acids and glycerides. The pancreatic juice will not function properly without the presence of the other two secretions. Thus protein digestion by trypsin has to be initiated by the conversion of a precursor – trypsinogen – into trypsin proper by another enzyme – enterokinase – liberated by the duodenal mucosa, while fat digestion is enormously helped by the bile, which emulsifies the globules to expose a far greater area to the action of lipase.

Liver and bile

The liver is composed of myriads of lobules, each with a dual supply of ordinary arterial blood from the hepatic artery and of blood from the intestine, containing products of digestion via branches of the portal vein. Bile is continuously secreted by these lobules to a total of some 850 ml a day; as it is only required at intervals it is stored in the gall bladder, where it is concentrated by reabsorption of water and released when required.

Bile is not only a digestive juice but also a means of ridding the body of the breakdown products of haemoglobin. It is a slightly alkaline, viscid green fluid. The *bile pigments* are breakdown products of effete red cells, which are disintegrated in the liver and spleen. These pigments, altered and mixed with undigested food, give the faeces their characteristic brown colour. In *jaundice*, due to obstruction of the bile duct which prevents the pigments entering the bowel, the faeces are pale and clay-coloured, while the dammed-up bile overflows into the circulation and stains the skin its yellow. (Jaundice may also result from disease of the liver itself, as in infective hepatitis.)

The *bile salts* are the sodium compounds of certain complex acids and are essential in fat digestion. The bile contains no enzymes of its own, but its salts emulsify the fat of the food to facilitate its digestion by pancreatic lipase. The fats, split into glycerides and fatty acids, enter the bowel wall together with the bile salts; here, in the individual cells, globules of neutral fat are reformed and absorbed, while the salts return to the liver to be used again.

The muscular gall bladder is normally full between meals. Entry of food into the duodenum relaxes the sphincter muscle guarding the opening of the common bile duct into the bowel, and the gall bladder contracts and discharges bile.

The large intestine

The *large intestine* is little concerned with digestion or absorption; most of this has already been accomplished in the small bowel. However, there is one absorptive function of importance – the taking-up of water from the faeces. These enter the colon in a fluid state; there, they lose four-fifths of their water content, becoming solid when the lower (left) half of the colon is reached. Much of the excreta is composed of the enormous numbers of bacteria found in the large intestine; in the stomach and the upper small intestine the contents are relatively sterile. As a result, an abdominal wound that penetrates the large intestine may liberate bacteria into the peritoneal cavity, setting up *peritonitis*. The *faeces* are derived mainly from substances formed within the intestines. They contain 70% of water, some 15% of minerals, such as calcium phosphate and iron salts, and much nitrogenous material, mainly from dead bacteria. An increased cellulose (fibre) content in the food increases the bulk of the faeces and stimulates bowel passage. The faeces also contain shreds of mucous membrane and are passed together with gaseous *flatus*, mostly methane.

Movements in the digestive tract

The general principle in the bowel is that food is propelled onwards and downwards, by *peristalsis*. This is a rhythmical wave of contraction travelling down a segment or loop of bowel at about 2·5 cm a second, presenting a local ring-like spasm at any particular point, preceded and succeeded by a wave of relaxation. Peristalsis is not dependent on central control, for it is brought about by a nerve plexus within the muscle coats of the intestine. It is, however, subject to modification by the central nervous system.

In the small intestine there is another type of movement, *seg-*

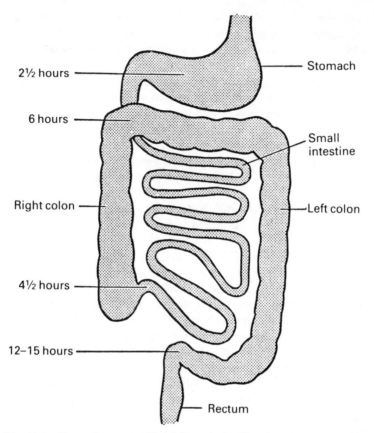

Fig. 13.2 Rate of passage of food through the digestive tract

mentation, in which a length of bowel is demarcated into sausage-shaped segments which reform from time to time. This does not propel food along the digestive tract. Finally, there are *pendular movements*, waves of contraction passing backwards and forwards, churning the contained food.

Regular peristalsis does not occur in the colon, where the problem is one of storage of the faeces until a new meal makes it necessary to move on its contents. Entry of food into the stomach excites as the *gastrocolic reflex*, a mass contraction that transfers the faeces from left (descending) colon to rectum, preliminary to *defaecation*. Defaecation consists of expulsion of the faeces outside the body, and the rectum normally remains empty until just before the act. Although the act is reflex, it is under conscious initiation and control; the urge passes off if the call is not obeyed, and the rectum relaxes. In the chronically constipated the rectum always contains some faeces, and loses its power of responding effectively to distension. During the act, the levator ani muscle (Fig. 8.21) lifts the anal canal over the faecal mass, while evacuation is aided by voluntary contraction of the diaphragm and abdominal muscles to raise the intra-abdominal pressure. The external opening of the anal canal is normally closed by its two guarding circular muscles or sphincters.

How long does it take for the food to arrive at different stages in its passage through the bowel? A 'meal' of barium sulphate, which is opaque to X-rays, can be watched during its passage along the alimentary canal (Fig. 13.2). The bulk of the stomach contents have passed on after three hours, though a small residue may persist for an hour or two. Food arrives at the lower end of the small bowel and enters the large intestine after four to four and a half hours; fills the right half of the colon by six hours; and occupies the left half, ready for expulsion into the rectum, at twelve to eighteen hours. These are average times and vary considerably; food may be detained in the large bowel for several days.

Vomiting is a reflex action initiated by many different stimuli. These include irritation of the stomach lining by food, alcohol or drugs; touching the palate or pharynx; distension or irritation of the stomach or bowel; stimuli from the vestibular apparatus of the inner ear (p. 337), as in seasickness; and severe pain or emotional disturbance.

The vomiting centre is in the medulla of the hindbrain (pp. 319, 320). There is a violent contraction of the muscles of the abdominal wall which expels the gastric contents through a relaxed oesophagus. Vomiting is usually accompanied by nausea, sweating, salivation and an accelerated heart rate.

14

Absorption, Utilisation and Storage of Digested Food

Absorption

Digestion transforms complex insoluble food substances into simpler soluble particles capable of entering the circulation. Their transference from the bowel to the blood or lymph is known as *absorption*. Absorption is dependent on physical factors such as concentrations and osmotic pressure, but most nutrients are actively transported and only water and alcohol cross the intestinal cells by simple diffusion.

Most nutrients are absorbed in the small intestine, the lining of which is not smooth, but a velvety pile of innumerable hair-like processes, or *villi*. These enormously increase the area available for absorption. Each villus (Fig. 14.1) is richly supplied with blood capillaries, and is traversed by a central lymphatic channel, or *lacteal*, for carrying digested fat globules to the lymphatic trunks of the body. In the fasting state the villi are at rest, but during absorption there is a constant pumping motion which squeezes absorbed materials into the circulation.

Water

A lot of water enters the digestive tract daily. In addition to this *external load*, say 1500 ml, an *internal load* is contained in the digestive juices: saliva 1500 ml, gastric juice 3000 ml, bile 500 ml and pancreatic juice 200 ml – a total of 8500 ml. This is all absorbed from the bowel, most from the small intestine and the remainder in the large intestine, where the faeces become more solid. Most water

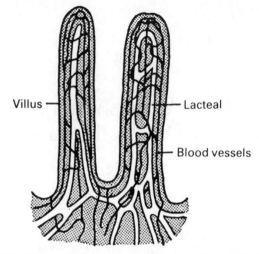

Villus — — Lacteal

— Blood vessels

Fig. 14.1   Villi of the small intestine with their lacteals and blood vessels

enters the portal venous blood and only a little enters the lymphatics. The transference of water (and dissolved electrolytes) across
the bowel wall into the vessels of the villi is not fully explicable in
terms of osmosis; it is influenced by the selective activity, or ionic
pumping action, of the intestinal cells themselves. The absorbed
water burdens the circulation and kidneys; hence the importance in
heart and kidney disorder of restricting intake.

Vitamins and minerals
The water-soluble *vitamins* are simple, soluble molecules and are
readily absorbed, although the absorption by the ileum of vitamin
B_{12} depends on an intrinsic factor secreted by the stomach. Most of
the *calcium* ingested is lost in the faeces, but some is absorbed to
supply the needs of the skeleton, more in growing children. This
absorption is dependent on vitamin D; if this is defective, or the
calcium intake is inadequate, rickets will develop in children and the
equivalent bone softening – called osteomalacia – in adults. The
absorption of *iron* is related to requirements; excessive intake leads
to damaging deposits in the liver and tissues.

Carbohydrates

Carbohydrates are broken down by digestion into simple monosaccharides and disaccharides, and the latter are changed into the monosaccharide form by the cells of the bowel wall, followed by active transport across the cell into the portal blood. There are specific enzymes for the different sugars, and if these are absent, disaccharides may pass unchanged into the colon and cause diarrhoea; some infants possess no lactase enzyme to digest lactose and cannot tolerate milk.

Fat

Very little fat escapes digestion and it is absorbed only in the small bowel. Fat leaves the stomach as large globules and is finely emulsified in the duodenum by the bile salts, and rendered easier to break down (hydrolysed) by pancreatic lipase into fatty acids and glycerides. These enter the cells of the bowel wall as a micro-emulsion and recombine within the cells of the villi into small globules, or *chylomicrons*, of neutral fat, which are transferred to the lacteals and thence to the lymphatics. After a fatty meal, these channels and the thoracic duct are distended by a milky fluid, the *chyle*, rendered opalescent by fat particles. The thoracic duct empties into the great veins of the neck, so that the blood serum may contain obvious fat a couple of hours after a meal. The process is accelerated by exercise.

Proteins

Under the influence of the pepsin of the gastric juice and the trypsin of the pancreatic secretion, proteins are broken down into amino-acids and peptides. The amino-acids are readily absorbed into the general circulation and little of the dietary nitrogen is lost with the faeces.

Utilisation (metabolism) of absorbed material

Protein metabolism

The place of proteins in the diet was dealt with on p. 26. We shall now consider their fate within the body, beginning with a look at their chemical composition. They are large-molecule, nitrogen-containing compounds with many differing functions. Some, like the keratin of hair and skin, are insoluble and inert; others fulfil the mechanical requirements of certain tissues – muscle and connective; some are soluble and used for transporting oxygen, minerals and hormones; and every enzyme is a protein. Man is incapable of directly utilising inorganic nitrogen, and depends on ingesting the protein of other life forms, which he degrades into their simple amino-acid constituents to be recombined in human patterns.

The amino-acids are the protein currency of the body – small, universally recognised, freely interchangeable and capable of being built up into the complex proteins of particular tissues.

All proteins can be hydrolysed to their constituent amino-acids, by boiling or by treatment with a series of enzymes. *Simple proteins* so treated yield amino-acids only. *Conjugated proteins* yield amino-acids, plus a linked prosthetic group that confers a specific character on that molecule. About twenty amino-acids are obtainable from protein hydrolysis, and their basic chemical formula is:

$$R\!-\!CH\!-\!COOH$$
$$|$$
$$NH_2$$

where, in the simplest form, $R = H$ (glycine).

In more complex forms, R may be an aliphatic chain or a cyclic radical. All amino-acids are optically active, but only their laevorotatory forms are yielded by hydrolysis. They can also combine with both acids and bases, and so they can act as neutralising buffers. Soluble proteins can be isolated and identified by chromatography, which depends on their differing rates of diffusion along moist filter-paper, yielding separate stainable protein 'spots', and also by electrophoresis, which depends on differential migration in filter-paper to one or other of a pair of electrodes.

The amino-acids in a protein are linked by peptide bonds, $-CO\!-\!NH\!-$, so that the basic structure of a protein molecule may be represented as:

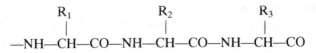

where R represents specific amino-acid radicals. During hydrolysis

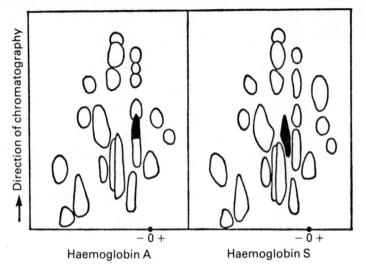

Direction of chromatography

$- 0 +$ $- 0 +$

Haemoglobin A   Haemoglobin S

Fig. 14.2 Fingerprints of tryptic hydrolysates of haemoglobins A and S. 0, point of application of hydrolysate; + −, direction of electric field. The peptides which differ in behaviour and composition are marked black; the one containing valine (in HbS) has moved slightly further towards the cathode (From *A Companion to Medical Studies, Vol. 1*)

polypeptides are formed, which are intermediate between proteins and amino-acids.

The molecular weights of proteins vary from a few thousand to over a million, i.e. the simplest protein molecule contains over a hundred amino-acid residues and several thousand polypeptides. Although there are only some twenty constituent amino-acids, there are many more different kinds of protein because proteins differ according to their constituent acids and how these are linked. Proteins may be 'finger-printed' by hydrolysing them to mixtures of peptides and separating these by electrophoresis, followed by paper chromatography, which yields a specific pattern for each (Fig. 14.1).

The shape of the protein molecule may be compact and globular in the case of the soluble proteins or rigid and elongated in the case of the fibrous proteins, often insoluble. They are frequently folded in helical manner, so identifiable by X-ray crystallography (Fig. 14.3). Proteins may be *denatured* by heat, acids or alkalis, which weaken the internal bonds of the molecule.

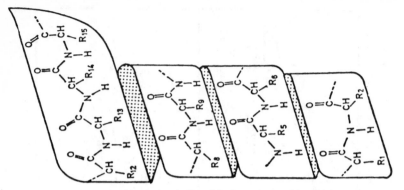

Fig. 14.3   The right-handed α-helix with the polypeptide structure drawn upon it (Modified from Haggis, G., *et al.*, *Introduction to Molecular Biology*, Longmans)

Proteins in nutrition

Energy is obtainable from all three basic foods, but only proteins will serve the needs of growth and tissue repair, i.e. the ultimate fate of the constituent amino-acids is partly to enter into living proto-plasm and partly to serve as a source of energy. In this latter function, protein is on a par with fats and carbohydrates; in the former, it is irreplaceable.

There is great variation in the capacities of proteins to support life. Young rats fed on maize protein (zein) lose weight because zein contains neither lysine nor tryptophane. If these are given as a supplement, growth resumes normally, i.e. these acids cannot be synthesized in the body. Thus they are known as *essential* amino-acids and must be supplied in the diet. For man, the essential amino-acids are: isoleucine, leucine, lysine, phenylalanine, tyro-sine, histidine (infants only), methionine, cystine, threonine, tryp-tophane and valine. Hence the importance of a well-balanced diet; and hence those living on a single staple risk developing protein deficiency. Note that the description 'essential' refers only to amino-acids which are indispensable in the diet because they cannot be synthesized in the body; the so-called 'non-essential' amino-acids which can be synthesized are nonetheless vital.

The *biological value*, or net protein utilisation, is:

$$\frac{\text{nitrogen retained in body}}{\text{effective dietary intake of nitrogen}}$$

a factor that can be calculated as:

$$\frac{\text{dietary } N_2 \text{ intake} - \text{faecal loss of } N_2 - \text{loss of } N_2 \text{ in urine}}{\text{dietary } N_2 \text{ intake} - \text{faecal loss of } N_2}$$

The biological value of a reference protein such as milk is standardised at 100%, so the value is less for less utilized proteins. Any constituent of the diet can be evaluated for its biological value and so it is possible to calculate a *protein score* for different foodstuffs.

The nitrogen required in the daily intake is equal to the daily nitrogen loss, plus that required for growth or other demands, such as pregnancy or lactation. The nitrogen lost with the faeces and from the body surface is largely independent of intake, but the loss in the urine varies with intake. The *obligatory nitrogen loss* is that lost daily in the faeces and urine, and from the body surface while on a protein-free intake. This obligatory loss is a measure of protein breakdown, or catabolism, and is increased by stresses such as infection, injury, pregnancy, lactation, fever, convalescence and emotional strains. Because of their growth requirements and higher metabolic rate, infants require more protein per kilogramme of body weight.

There is a relation between protein nutrition and calorie supply. Protein is utilised most effectively when the bulk of energy requirements is met by other foodstuffs, particularly carbohydrates. If the carbohydrate intake is inadequate, some protein must be diverted to supply energy and there is a risk of protein deficiency. If calorie needs are met by a diet poor in protein, it is impossible to eat enough food to supply the intrinsic protein need. It is possible to supply all the calorie needs of a growing child with deproteinised milk; but protein deficiency will develop unless a supplement is given in the form of dried, skimmed milk.

Protein transformations in metabolism
Proteins take part in many chemical processes; many different enzymes are involved and each protein is metabolised in a specific manner. The proteins of the tissues are not static, but constantly receive fresh amino-acid constituents from the blood and shed their

old ones. The end-products of amino-acid degradation are CO_2 and water, plus nitrogenous compounds such as urea and creatinine, which are lost in the urine. Amino-acids are present in the blood plasma and enter the interstitial (extracellular) fluid, to be taken up by tissue cells; the amino-acid concentration within the cells may be five to ten times the concentration in the extracellular fluid. Little amino-acid is normally excreted *as such* in the urine because of selective reabsorption back into the blood in the renal tubules (p. 299). Amino-acids are degraded by a process of *deamination*, an oxidative process in which the nitrogen is converted to ammonia. This is converted in the liver to urea, which is excreted by the kidneys. Most of the amino-acid carbon is eventually lost as CO_2; but a significant amount of glucose is formed from the split protein. The sulphur of certain amino-acids (cysteine and methionine) is excreted in the urine but some becomes linked to polysaccharides and forms the ground substance of cartilage.

Protein degradation is mainly a process of hydrolysis, liberating amino-acids, some of which are utilised to replace effete intracellular proteins. Some idea of the length of life of protein molecules is obtainable by tagging them with identifiable isotopes; human serum albumin is replaced at the rate of 10% a day. The rate of turnover of protein in the body as a whole is some 250 g a day, i.e. some 2½% of the total body protein.

This leads to the concept of *nitrogen balance* as between daily dietary intake and loss by excretion, mainly in the urine; in the steady state these are in equilibrium. But on a protein-free diet the balance becomes negative and the urinary output falls to the basal obligatory loss; the weight and protein content of the liver and other viscera fall, and cell division and replacement slow down. Return to a normal diet reverses these changes, so there is a positive nitrogen balance until equilibrium is restored.

Fat metabolism

Fats are compact, readily mobilised, rich stores of energy. They are insoluble in water. The main groups are:

1　Fatty acids;
2　Triglycerides and waxes;
3　Phospholipids;

4   Cholesterol and its derivatives;
5   Vitamins A, D, E and K.

The *fatty acids* are unbranched aliphatic chains with one terminal carboxyl (COOH) group. The simplest form is acetic acid, CH_3COOH, and the type formula of the complex varieties is

$$CH_3\text{---}(CH_2)_n\text{---}CH_2COOH$$

Fatty acids may be saturated or unsaturated, the latter having an unsatisfied capacity to accept hydrogen; commercially, hydrogenation is used to convert unsaturated vegetable oils to solid fats. Unsaturated fatty acids may be valuable in preventing arterial disease.

The *triglycerides* comprise the bulk of dietary fat, and the basic model is an ester of glycerol with three fatty acid chains. All fats are triglycerides, differing only in the nature and amount of the fatty acids esterified to the glycerol base. Referring to the number of carbon atoms in the fatty acid chain, human fat consists mainly of C_{16} and C_{18} acids, saturated and unsaturated. Fish oils have a high proportion of polyunsaturated fatty acids and are liquid. Saturated lard and suet are solid at room temperature.

The *phospholipids* contain phosphorus and sometimes nitrogen. They are important components of cell membranes, plasma, and the nervous system.

Cholesterol is a very important compound of a basic ring structure:

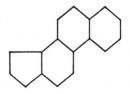

Some comes in the diet; more is synthesized in the body, notably in the liver. It is an example of a class of lipids known as *steroids*. These are most important compounds and include the hormones of the adrenal and sex glands, vitamin D and the bile acids.

The fate of absorbed fat (lipids)
About 100 g of fat is absorbed daily, most as triglycerides. These enter the duodenum as a crude emulsion, are finely divided by the bile salts, and then split by pancreatic lipase to fatty acids, mono-glycerides and some glycerol. These are not water-soluble and their transfer across the bowel lining is facilitated by bile salts, which form them into tiny aggregates which enter the villi. Within the villus cells they recombine, though not into the original tri-glycerides.

They enter the lymph vessels in small particles known as *chylo-microns*. These enter the central lacteal of each villus and travel up the thoracic duct.

Lipids are transported in the plasma as *lipoproteins*, large mole-cules of triglycerides embedded in cholesterol, phospholipids and protein. These convey the triglycerides to the tissues, where they are separated for local use, while the conveyor system shuttles back to the liver to pick up more triglycerides. The *serum lipids* consist of:

1 Cholesterol, in amounts of 150–250 mg/100 ml in youth, rising to 300 mg or more in later life;
2 Phospholipids in amounts of 150–300 mg;
3 Triglycerides in amounts of 50–300 mg in the fasting state, but reaching much higher levels after a fatty meal.

The *cholesterol* goes to the liver, where it is converted into bile salts and excreted into the bowel. However, in atherosclerosis, common in middle-aged and elderly men but less common in women, cholesterol is deposited in the lining of the arteries and may obstruct them.

In the tissues, the arriving lipoproteins are cleared from the plasma and the triglycerides are split by a lipase enzyme. The fatty acids enter the fatty tissue and are reconverted to triglycerides; or they enter muscle to supply energy by oxidation. The freed glycerol returns to the liver. Thus part of the transported serum fatty acids is stored in the body fat and part is burnt to supply the energy for muscular contraction. During exertion the fatty acid content of the plasma rises and more is taken up by muscle. This mechanism is less important if there is ample free glucose available. In brief exertion the initial source of muscle energy is the glycogen stored in the muscle cells, which is split to yield glucose. But after some minutes the muscles begin to take up free fatty acids from the blood (see p. 59).

The energy store represented by the triglyceride content of adipose tissue is available when required, the glycerides being split back to glycerol and fatty acids which enter the bloodstream, from which the acids are taken up by the organs and oxidised to yield energy.

The carbohydrates of the diet may also contribute to the fat store. The glucose molecule is broken down and the carbon atoms re-assembled in long, fatty acid chains.

Cholesterol is synthesized in the body tissues, and that present in the blood is partly free and partly esterified. The *phospholipids* are essential to the integrity of cell membranes, which control the transfer of various small molecules, nutrients, wastes and water.

Fatty acids and carbohydrates

Fatty acids and carbohydrates are broken down in the tissues to yield CO_2, H_2O and energy. In the resting state most of the energy requirement is obtained from the fatty acids, less from glucose, and even less from amino-acids, ketones and glycerol. But this changes with eating or exercise. After a carbohydrate meal the respiratory quotient moves nearer to 1, i.e. glucose rather than fat is being used to provide energy. During violent exertion, the main initial source of energy comes from the glycogen store of muscle, but if the exertion continues, the muscles begin to utilise free fatty acids from the blood, and the respiratory quotient falls. During fasting the liver content of glycogen is depleted, the blood level of glucose falls, less is used in the tissues and the metabolism turns over to fat. However, the brain always prefers glucose for energy transformations.

Ketosis

If insufficient glucose is available, because of fasting or prolonged exertion, energy is obtained predominantly by the mobilisation of fat stores; this metabolism yields *ketone bodies*, such as acetoacetate, and this lowers the pH of the blood – a condition of *acidosis*. Ketone bodies also appear in the urine. Similar changes may occur in uncontrolled diabetes.

The liver plays a cardinal role in fat metabolism; it normally contains some 5% of fat, but more may appear in disease such as fatty degeneration, starvation, and certain forms of chemical poisoning.

Fatty (adipose) tissue

The basic fat is widely distributed and makes up some 10% of the body weight. About half is under the skin, most of the remainder in great sheets within the abdominal cavity, or as a packing round the internal organs, a little in the intermuscular planes. There is none in the cranial cavity. A healthy woman has about twice as much fat as a healthy man, but the quantity present in any one person depends on nutritional and constitutional factors. Its distribution is affected by the sex hormones; there is more around the hips and thighs in women, more in the upper half of the body in men.

Fat insulates the body against temperature changes, and buffers the internal organs; it is an electrical insulator around the fibres of nerves. Microscopically, fat is divided into lobules by fibrous partitions. The fat cells are connective tissue cells that have become deformed by a central fat globule, which pushes the nucleus to one side.

Chemically, fat stores fatty acids in the form of triglycerides as a source of energy when required by changes in nutrition and environment. It consists of some 85% fat (mostly as neutral triglycerides, though there are also some free fatty acids, glycerol, cholesterol and phospholipids), 12% of water and 3% of protein. But adipose tissue is not a passive storehouse of triglycerides; it is an entrepôt where a continuous arrival of new material is broken down and resynthesized to pass out to the tissues. There is a constant turnover. Fat can be formed even on a fat-free diet from glucose in the blood, synthesis occurring in the liver and other tissues.

Fat is added to the reserve when the food exceeds energy needs. Different animals have characteristic fats, e.g. human fat is liquid at body temperature, mutton fat solid. This depends on the capacity of the body to modify fat taken in the food.

Fear, cold and other stresses mobilise fatty acids, and this response is mediated partly by adrenal hormones and partly by the autonomic nervous system.

Carbohydrate metabolism

Glucose is manufactured from CO_2 and water by photosynthesis in the green plant. It is rarely found as such in nature but occurs in polymer form; and starch is a polymer in cereals that man can digest; he is barely able to utilise the cellulose polymer of plants.

The basic building block of carbohydrates is CH_2O, and their general formula is $(CH_2O)n$. The commonest carbohydrates are monosaccharides, where n = 5 or 6, pentoses and hexoses. These are soluble and sweet. Disaccharides, such as maltose, are made by linkage of two monosaccharides. Ascorbic acid (vitamin C) is a hexose derivative.

Saccharides exist as stereoisomers, i.e. the same compound has different asymmetrical spatial arrangements of atoms in the molecule. The number of possible isomers may be large; there are sixteen for a hexose. Now, although the chemical analysis of isomers is identical, they differ in their physical and biological properties. All rotate the plane of polarised light to right (dextrorotatory) or left (laevorotatory). However, the body can use only one isomer of any compound, and the others, though chemically identical, exert little or no physiological action. Thus l-ascorbic acid is active against scurvy, whereas d-ascorbic acid is not; and l-thyroxine is a stimulator of general metabolism, whereas d-thyroxine is only a third as effective. When such a compound is manufactured in the laboratory, it consists of a *racemic mixture* of equal proportions of both isomers which is optically inactive, since the rotations produced by the constituents cancel out, and has only half the biological activity of the natural isomer.

In the metabolism of glucose, a basic reaction is with adenosine triphosphate or ATP (a nitrogen-containing purine body) that yields glucose-6-phosphate and adenosine diphosphate (ADP). This reaction is catalysed by an enzyme called hexokinase. Glucose-6-phosphate is the basic carbohydrate currency of the body, just as the amino-acids are the basic protein currency. It is the starting-point for glucose/carbon metabolism, and there are many pathways and intermediate products, in which organic phosphates and phosphatase enzymes are prominent. The basic final degradation of glucose to yield energy is to pyruvate (pyruvic acid is $CH_3(CO)COOH$, which is then split in the tissue cells to CO_2 and water).

Glycogen is found everywhere but mostly in liver and muscle. It is a polymer of glucose with a molecular weight of several million.

Lactose is the disaccharide of milk, synthesized from the monosaccharides glucose and galactose only in the breast. In the infant, it is hydrolysed in the bowel wall to glucose and galactose, and the latter is converted to glucose. If the last sequence is impossible because of congenital lack of the appropriate enzyme, galactose accumulates and may cause stunting or mental defect; the only cure is a diet containing no lactose.

The liver is the only organ that can form free glucose; and the blood sugar level cannot be permitted to fall below a certain threshold without grave consequences. This is because there is a continual demand for glucose from the blood by the various organs, particularly the brain. Therefore, even in starvation or on a carbohydrate-free diet, the liver must manufacture glucose for as long as possible.

In discussing protein metabolism, we used the simile of a reservoir constantly emptied and refilled. This applies equally to the sugar of the blood. There is a constant intake from the food and from the liver glycogen; and there is a constant output into the tissues, where some is burnt for energy and some stored as a glycogen reserve. The blood sugar is thus in equilibrium with that of the tissues, and is a very efficient mechanism; maintaining its concentration fairly constant. The blood sugar rises somewhat after a meal but is back to normal within two hours. This regulation is so effective that it remains much the same in starvation as after a heavy meal; only exceptionally does the level rise so much that some sugar spills over into the urine.

How does this mechanism work? The controlling agent is *insulin*, an internal secretion or hormone of the pancreas. Insulin is essential to the utilisation of sugar by the tissues and is normally formed by the pancreas in amounts just sufficient to deal with sugar added to the circulation from the food; it keeps the blood level down to normal by enabling the tissues to use this added sugar. Without insulin, the tissues cannot properly burn glucose, sugar is dammed up in the blood and spills into the urine. The loss of this source of energy results in relative starvation; also, fat metabolism is interfered with, and acidosis results. This is the condition of *diabetes*, due to deficiency in the secretion of insulin. The treatment depends on giving insulin artificially by injection, and on balancing the insulin given and the sugar in the food so that the blood level is kept normal. In minor cases, it may be possible to manage merely by reducing the sugar in the diet. We might expect that an excess of insulin would abnormally depress the blood sugar; this is so, and the consequences may be serious – acute anxiety, hunger, pallor and sweating, convulsions, even coma and death from hypoglycaemia. The same results occur in the rare condition of a tumour of a cell-island in the pancreas in which insulin is formed; life may depend on its location and removal. Slight overproduction of insulin in normal individuals may cause recurrent restlessness or even mental disturbance; this is rapidly alleviated by eating sugar or sweets.

The liver is the principal store of glycogen, and constantly converts it into glucose and vice versa; it also forms glycogen from protein and fat. Just as insulin regulates blood sugar outflow by enabling the tissues to burn it, so the liver regulates inflow by adding from its glycogen stores. This liver store is a reserve against starvation, and quickly converted into soluble glucose and passed into the circulation; of all the body's reserve, this is most rapidly available and easily used. In its mobilisation *adrenaline*, the internal secretion of the adrenal glands, plays a part; in fear, struggle and cold, adrenaline acts as a potent messenger to the liver, and sugar is rapidly poured into the blood as an immediate source of energy.

The muscles are the other main stores of glycogen, to supply their energy demands in contraction. Then they convert the glycogen not to glucose but into lactic acid, which enters the blood and is reconverted to glycogen in the liver or the muscles again. More glycogen is stored in the muscles than in the liver, but is kept for muscular exertion and is not readily available for general use.

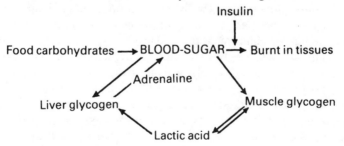

The liver in metabolism
The liver is of cardinal importance for metabolism. It plays an important role in the synthesis, storage, degradation and inter-

conversion of protein, fat and carbohydrate. Bile formation, the synthesis of urea from ammonia radicles split off proteins, the metabolism of drugs, hormones and toxins, are all the exclusive province of the organ. If the liver ceases to function normally, serious illness – even death – may follow. Yet, it has a huge functional reserve and there may be no obvious disturbance even when it is extensively diseased.

Because the liver is important in the synthesis of proteins, the interconversion of amino-acids and the production of urea as an end-product of protein metabolism, in liver failure the amino-acid content of the blood rises and there is a fall in urine urea and blood urea. Protein is formed in the liver partly from dietary amino-acids and partly from amino-acids derived from tissue breakdown. Protein breakdown occurs by deamination of amino-acids, and the potentially toxic ammonia so yielded is converted into non-toxic urea. Other tissues, particularly the kidney, also form some ammonia, and this is brought to the liver for conversion.

The liver plays a special part in the formation of the *plasma proteins*. These include: albumin, the most abundant protein in the blood plasma; alphaglobulins and betaglobulins in smaller amounts; proteins concerned with blood coagulation; and certain enzymes. The plasma proteins have a short life and are constantly renewed. Their analysis may be effected by electrophoresis, which yields a definite pattern (Fig. 14.4), which is altered in disease. If the diet is poor in protein, the plasma proteins fall, and their osmotic action of holding water is reduced; so in famine conditions, fluid will seep out of the blood to cause oedema of the subcutaneous tissues.

We have already seen the liver's function in maintaining blood glucose, and the part played by hormones – insulin and adrenaline – in this mechanism, at rest and in exercise. The liver normally stores about 100 g of glycogen, immediately available to maintain blood glucose levels. But it also actively manufactures glucose: from glycerol formed in the breakdown of fats, from monosaccharides, and from the breakdown products of amino-acids. This *gluconeogenesis* is promoted by certain hormones: the adrenal hormone cortisone, thyroxine and a pancreatic hormone, glucagon.

The liver both synthesizes lipids, arriving from fat depots or from the food, and breaks them down, the long chain fatty acids being oxidised to water and CO_2. If this breakdown is impaired, ketone

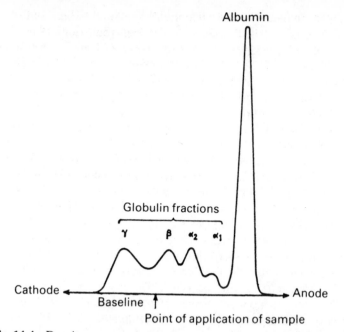

Fig. 14.4 Densitometer tracing of an electrophoretic separation of normal serum proteins (From *A Companion to Medical Studies, Vol. 1*)

bodies are formed – acetoacetic acid, hydroxybutyric acid and acetone – giving rise to ketosis or acidosis. Ketosis is an index that predominantly fat metabolism is going on, as in starvation, in prolonged fever, after long exertion or in uncontrolled diabetes.

Cholesterol is synthesized in the liver (and elsewhere) and excreted in the bile to be reabsorbed from the intestine. The liver is also important in the formation of lipoproteins. Nine-tenths of ingested alcohol is oxidised in the liver. The bile salts are formed in the liver from cholesterol.

Of the bile pigments, the most important, *bilirubin*, is formed by breakdown of the haemoglobin of the red blood corpuscles. This is brought to the liver, conjugated in the diglucuronide form, and excreted in the bile to enter the bowel. There bacteria convert it into *stercobilin*; some of this serves to colour the faeces, while some returns to the liver and circulates again and again between bowel and liver or is excreted by the kidneys as *urobilin*, the yellow colour of urine. *Jaundice* occurs when the normal level of bilirubin in the plasma is exceeded. It may be due to overproduction of bilirubin, with accumulation in the plasma, disease of the liver, or obstruction to the outflow of bile from the liver. The first type occurs when red cells are destroyed in unusually large numbers, as in some anaemias or malaria; liver disease impairs the transport of bilirubin; and obstruction to outflow of bile, as from a stone or cancer in the bile ducts, dams back the pigment. In most forms of jaundice, bilirubin appears in the urine.

The liver also metabolises certain hormones, as well as many drugs and toxic agents; if it fails to do so, liver failure may result, one feature of which is fatty degeneration.

The liver is also the chief depot of the fat-soluble vitamins A, D, E and K, and holds enough of the first two to prevent deficiency disease for a long time, even if they are absent in the diet. Vitamin K is used for the synthesis of clotting factors. Water-soluble vitamins are also stored, and the liver's enzyme systems can convert dietary tryptophane into nicotinic acid (p. 230)

There are a number of tests of *liver function*. These include measurement of plasma bilirubin or serum albumin, the first being raised and the second lowered in liver inefficiency; a galactose-tolerance test, designed to assess the organ's ability to convert galactose to glucose; and a test for escaped liver enzymes in the serum.

15

Blood, Lymph and the Reticuloendothelial System

Blood

The blood is a complex fluid pervading all parts of the body – except cartilage and the cornea – that conveys food and oxygen to the tissues, and carries away wastes. It varies between the purple of venous blood and the bright red in the arteries, according to its oxygenation. Microscopic examination shows myriads of *corpuscles* of different kinds suspended in the fluid component, or *plasma*. Corpuscles and plasma occupy an approximately equal volume and can be separated by standing or centrifuging. The blood cells consist of red cells or *erythrocytes*, white cells or *leucocytes*, and platelets or *thrombocytes*. In centrifuged blood the deposit consists of red cells; above is a thin layer of white cells and platelets, the *buffy coat*; and above this is the clear, yellowish plasma.

The plasma is a clear, pale yellow fluid containing some 10% of solids – mostly protein – together with salts, particularly sodium chloride. There are also sodium bicarbonate, phosphates, potassium, and representative metabolites of the different foodstuffs – glucose, urea, amino-acids, fatty acids, and so on. The proteins of the plasma are coagulable, and one, fibrinogen, is concerned exclusively with clotting. The albumin and globulin are responsible for its considerable osmotic pressure.

Blood cells

The *red corpuscles* are by far the most numerous, some five and a half million per cubic millimetre. They are cells that have lost their nuclei – biconcave discs, narrow-waisted in profile, which are envelopes for the respiratory pigment, haemoglobin. The splitting of haemoglobin into innumerable little packets vastly increases the surface area for oxygen interchange; the total area of the red cells is 1000 to 2000 times that of the body. They are sensitive to changes in the osmotic pressure of the plasma. An increase in osmotic pressure shrivels them by withdrawing water, while they swell up and burst in a weaker solution, their pigment is freed and the blood is *haemolysed*.

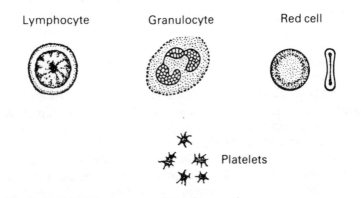

Lymphocyte   Granulocyte   Red cell

Platelets

Fig. 15.1   The different types of blood corpuscles

Before birth the red cells are formed in the bone marrow, liver, spleen and lymph glands. After birth the marrow is their only source; in early life the bones are full of red marrow, but later this retreats to the bone ends, leaving the shaft occupied by yellow fatty marrow. The survival time of the adult red cell is some 120 days, after which it fragments and is absorbed. The corpuscles develop from specialised cells which lose their nuclei; an intensive demand, in response to haemorrhage or anaemia, results in the temporary appearance of nucleated forms.

If blood treated with an anticoagulant is allowed to stand, the red cells sediment slowly, into clumps. The *rate of sedimentation* is influenced by the composition of the plasma proteins; it is often increased in disease and is an index of the activity of a disease process.

Haemoglobin consists of a protein – *globin* – linked to a chemical grouping – *haem* – containing iron, responsible for its colour. Its value is the ease with which it combines loosely with oxygen in the lungs to form bright red oxyhaemoglobin and with which this oxygen is given up to the tissues, leaving purple reduced haemoglobin. (Carbon dioxide is dissolved in the plasma.)

Many factors are essential for the maturation of red cells and for providing the right amount of haemoglobin, and lack of any one may cause anaemia. Amino-acids are required for the synthesis of globin, and iron is essential because of its central position in the haemoglobin molecule. Women need more iron than men because of blood loss in menstruation, and the growing child proportionately more than the adult. So, iron-deficiency anaemia is common in women and children but uncommon in men unless they are losing blood from a stomach ulcer or a blood-destroying parasite. We mentioned the importance of vitamin B_{12} and folic acid, and the gastric intrinsic factor essential to the proper utilisation of B_{12} (p. 231); lack of any one of these is the cause of *pernicious anaemia*. Other essential factors include thyroid hormone, copper and vitamin C, and a kidney secretion called *erythropoietin*.

In health a red cell of normal shape and size is termed *normocytic*, and, if it has a normal complement of haemoglobin, *normochromic*. Smaller or larger cells are microcytic or macrocytic; cells with reduced haemoglobin are hypochromic. The normal haemoglobin content of blood is 13·5–18·0 g/100 ml in men and 11·5–16·5 g/100 ml in women. *Anaemia* exists when the level is lower, either because there are fewer red cells or because their haemoglobin content is reduced, or both. These factors can be expressed numerically by relating haemoglobin and red cell count to *packed cell volume* (PCV), the proportion of blood volume occupied by packed red cells after centrifuging, some 40–55%. This gives the *mean cell volume* and the *mean cell haemoglobin concentration*:

$$MCV = \frac{PCV\%}{\text{Red cell count} \times 10^6 \, \text{mm}^3} \times 10$$

$$MCHC = \frac{\text{Haemoglobin (g/100 ml)}}{PCV\%} \times 100$$

Ordinary anaemia is due to lack of iron, where the MCV and MCHC are both reduced, i.e. the red cells are smaller and paler than usual and the haemoglobin may fall by half, i.e. *hypochromic anaemia*. In *pernicious anaemia*, however, the red cells, though fewer, are larger than normal and have their usual complement of haemoglobin. Anaemia causes pallor of the skin and mucous membranes, and breathlessness and easy fatiguability due to reduced oxygen-carrying capacity of the blood. The body can adjust to even severe anaemia if its onset is gradual; it may be possible to walk in comfort with a third of the normal haemoglobin if the cause is chronic, whereas a rapid fall to a much higher level may produce breathlessness. The listlessness and inertia of tropical races is often due to severe chronic anaemia produced by parasitic diseases or dietary deficiencies.

The infant at birth has (proportionately) more haemoglobin and a greater PCV than in later life. Sex differences are due to the effects of menstruation. The red cell content increases during exercise and emotional stress. In individuals at high altitudes, where barometric pressure and available oxygen are reduced, there is increased red cell production and haemoglobin content.

The white cells

There are three varieties: polymorphonuclears, lymphocytes and monocytes (Fig. 15.1). The total white count is 4000–10000/mm³. If this is markedly increased, as in infective disease, we speak of a *leucocytosis*; if decreased, as in toxic bone marrow failure, this is a *leucopenia*.

The *polymorphs* (granulocytes) have a segmented nucleus and are two-thirds of all the white cells, with a granular cytoplasm. 90% are neutrophils (which refers to their reaction to biological stains); a few are eosinophils or basophils, according to their affinity for acid or basic dyes.

The *lymphocyte* has a large, central basophil nucleus occupying most of the cell; nearly a third of white cells are lymphocytes. The *monocyte* is the largest and scarcest – some 3–4%. It has abundant cytoplasm and an indented or lobed nucleus.

The polymorphs are produced only in the red marrow and there is a large reserve. Their life span is 1–2 weeks. Polymorphs are lost in defence operations (see below), and are also shed into the body cavities.

The lymphocytes are formed in lymph nodes, the tonsils, the spleen, lymphoid patches in the small intestine and in the bone marrow. There is a large reserve in the lymphoid tissue. There are two types of lymphocytes, large and small.

The leucocytes are essential in preventing infection. The polymorphs destroy and remove foreign particles and bacteria, while the lymphocytes produce antibodies. The polymorphs also have the power of *phagocytosis*, or swallowing foreign material. At the site of injury or infection, they ooze through the walls of the blood capillaries and assemble to overcome the micro-organisms and remove waste material – including dead and damaged cells. If they are killed, their dead bodies form part of the pus. The monocytes and neutrophil polymorphs are the most mobile in the tissues, and are attracted to sites of damage by locally released substances. They flow round particles and engulf them in a streaming amoeboid movement. The other two types of polymorph are not so active, and lymphocytes are not phagocytic.

The *platelets* are tiny refractile bodies lacking a nucleus and number $150000-350000/mm^3$. They have the radiating processes of an asterisk in form. They are not true cells but fragments of bone marrow cells that have been shed into the circulation and assist the process of clotting.

Clotting
Blood clots a few minutes after being shed. The clot is a meshwork of a substance called *fibrin*, with blood cells trapped therein, derived from the fibrinogen protein of the plasma. This soon contracts, squeezing out a straw-coloured fluid, or *serum*, which is what remains of plasma after the loss of its fibrinogen. Serum and plasma are largely, but not entirely, identical.

Blood (Plasma + Corpuscles) → Clot (Fibrin + Corpuscles) + Serum

Clotting is a complex process requiring the presence of calcium salts. The fibrinogen-fibrin conversion is activated by an enzyme, thrombin, which is not present in normal blood but formed from an inactive precursor, prothrombin:

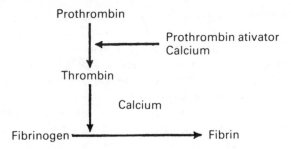

The process is complicated for there are many accessory factors. Clotting may take place in injured blood vessels or in shed blood in the tissues. When a blood vessel is divided, the injured end retracts; platelets are attracted to the damaged endothelium and aggregate to seal the cut end. After this initial phase of platelet plugging the secondary phase of clotting occurs. As the coagulation process must not spread back through the vascular tree there is a complementary process of fibrinolysis, in which the fibrin is dissolved, except where needed. Clotting is facilitated by substances derived from injured tissues, and if blood is collected carefully into a perfectly clean vessel it remains fluid for quite a considerable time. Other clotting agents include vitamin K, stored in the liver; the haemorrhagic tendencies in jaundice are due to its absence.

Blood may fail to clot normally in certain conditions. An essential clotting factor is constitutionally absent in 'bleeders' or haemophiliacs, victims of a disease transmitted to male descendants by female members of the family, themselves immune. Certain chemical substances delay or abolish coagulation. One of these is *heparin*, isolated from the liver; another is *dicoumarol*, analogous to the 'warfarin' used to poison rats. They have proved useful in surgery to prevent post-operative vein thrombosis and in the repair of injured blood vessels.

Blood groups

Human tissues and fluids are sensitive to foreign proteins and react against them. In solution this reaction causes their precipitation. Thus human plasma mixed with animal blood causes the red cells of the latter to clump together in sticky, granular masses in the process of *agglutination*. So animal blood is dangerous for transfusion: the agglutinated cells obstruct the recipient's blood vessels, while the debris after their destruction blocks the kidneys.

This sensitivity also exists between one individual and another, if they belong to different blood groups. Within the same group, transfusion is safe; outside this, only certain combinations can be used. Human blood groups are inherited with specific antigens (p. 268) on the surface of the red cells. There are four main groups: O, A, B and AB. Their compatibility in transfusion is shown in the following table. Note that it may be possible for a member of one group to donate blood to another, though fatal to receive blood from the latter.

Group	Can give blood to	Can receive blood from
O	O, A, B, AB	O
A	A, AB	O, A
B	B, AB	O, B
AB	AB	O, A, B, AB

Group O is a universal donor and found in 40% of the population. Group AB, safe for donation to its own group only, is a universal recipient, but occurs only in 2% of the population.

Blood groups are hereditary and are of value in cases of disputed paternity, though their evidence is only negative – i.e. the only certainty is that someone could *not* have been the father of a child. The racial distribution of blood groups differs in different parts of the world.

The red cells of some individuals possess a component analogous to one in the blood of the Rhesus monkey, the *rhesus factor*. Most people (about 85%) have this factor; they are Rh positive, and can marry and have children normally. But if an Rh positive man marries an Rh negative woman, and she becomes pregnant, the Rh factor inherited by the foetus from its father excites an antagonistic reaction in the mother. Substances formed in her blood enter the child's circulation, causing destruction of the latter's red corpuscles, so the infant may be born dead, or with serious anaemia or jaundice, or spastic because of haemorrhage in the brain. This disease is rare in a first pregnancy, but the risk increases with each successive pregnancy.

Blood volume and pH

The total volume of blood is 4–5 litres, varying with the size of the individual. After severe haemorrhage, the fluid loss is made good by entrance of water into the circulation from the reserves of extracellular fluid; there is dilution and a lower red cell count than normal. The reverse occurs in severe shock or dehydration; plasma leaks into the tissues, leaving a viscous concentration of corpuscles in the vessels. The blood is slightly alkaline, with a pH of 7·4. If we try to neutralise it with acid, more is required than would be expected. This is because there is an alkali reserve of sodium bicarbonate which, together with the plasma proteins, acts as a buffer, resisting any change in hydrogen-ion concentration. When this reserve is greatly reduced, as in ketosis, true acidosis develops, but this is exceedingly rare.

Lymph

Lymph is a pale yellow fluid, similar in composition to interstitial (extracellular) fluid, which clots on standing and contains lymphocytes. The lymphatic (lacteal) system of the bowel plays the main part in absorption of fat, so that lymph in the thoracic duct after a meal is milky and opalescent with fat globules. (For details of the *lymphatic system* see p. 269.)

Reticuloendothelial system

This system is not localised, but is a tissue widely dispersed throughout other tissues and organs. It consists (Fig. 15.2) of a meshwork of fine fibres supporting large, irregular cells and permeated by channels, or sinusoids, lined by an endothelium composed of these cells.

These cells are phagocytic and some wander actively in the tissues as scavengers. The monocytes of the blood form part of this system; and phagocytosis is an essential defence mechanism in combating invasion by bacteria or parasites. The body is liable to infective attack from three sources:

1 *Viruses*, like those of infantile paralysis and chickenpox, which are too small to be seen under the light microscope and traverse very fine filters;

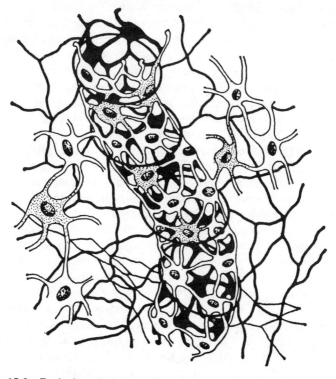

Fig. 15.2 Reticuloendothelium. Note the network of reticulin, the reticulum cells and the endothelial-like lining they form on the reticulin framework of the fluid channels or sinusoids (From *A Companion to Medical Studies, Vol. 1*)

2 *Bacteria*, visible microscopic organisms like those of plague and typhoid;
3 Larger parasites, which are single cells but still microscopic, the *protozoa* of malaria, syphilis and other diseases.

The primary defence is the outer coverings and internal linings of the body, the normally intact 'epithelial envelope'. The skin is a physical obstacle. The mucous membranes are more sensitive, but even here organisms become entangled in mucus or are wafted away by the hair-like processes (cilia) of the cells in certain regions (as in the air passages). The secretion of certain glands is bactericidal, as with the tears and saliva. An intact epithelium is normally proof against invasion, but the gonococcus and spirochaete, which cause gonorrhea and syphilis, can traverse the intact mucous membrane of the genital tract.

Once organisms have entered the tissues, the first defence lies in their being attacked by the phagocytes of the reticuloendothelial system that wander to the part and devour debris and organisms. Minor processes can be managed by the phagocytes locally available, but it is usually necessary to bring up reinforcements by increasing the local blood supply to provide many more phagocytes. All the capillaries dilate widely and the granulocytes make their way through their capillary walls to the site of infection, where they accumulate. When the infection is overcome and the debris removed, repairs are carried out by an overgrowth of fibrous tissue in the depths of the wound and an ingrowth of surface epithelium.

Although the reactions to injury and infection are similar, the latter requires many more phagocytes. In an infective focus still unresolved after forty-eight hours, *pus* (a fluid collection of dead phagocytes and bacteria) begins to form at the centre. However, pus is absent in wound healing, unless secondary infection develops.

This process of *inflammation* is marked by redness and warmth, due to increased vascularity, by swelling, due to exudation of fluid from the capillaries, and by pain due to stretching of nerve endings.

As well as this defence against invading organisms, there is a defence in depth against the foreign substances they produce. This operates throughout the body, is known as the *immune response*, and is a reaction to foreign substances, usually proteins, known as *antigens*. There is a synthesis of gammaglobulins in the serum, known as *antibodies*, which interact specifically with foreign antigens, neutralising their toxic effects, destroying bacteria or facilitating their phagocytosis. This defence is necessary because bacteria may affect the body as a whole, either by invading the blood and lymph streams, or by forming a toxin which circulates in the blood, or both. Some toxins, such as those of diphtheria and tetanus, are extremely virulent and may cause death, though the bacteria remain localised to the throat or a wound.

In most cases, the primary immune response by production of antibodies is not immediate; but, once formed, the individual is said to have an *acquired immunity*, since antibodies either persist or return very quickly to cope with subsequent attacks. Since the individual owes this immunity to having overcome an attack of the disease by manufacturing protective antibodies in his own tissues, this is described as *acquired active immunity*. Active immunisation may also be achieved by giving small doses of toxin or dead bacteria by inoculation or injection; the response is as great as after a real attack.

It is possible to produce actively acquired immunity against, say, tetanus by injecting a horse with increasing doses of toxin until the animal remains unaffected by a dose which previously would have been fatal. Its serum now contains a large amount of antitoxin. The injection of such serum into a man confers an *acquired passive immunity* – passive since his own tissues have played no part in elaborating the protective agent. This protection is only temporary, and is only of value in treating early disease, or in trying to forestall it. It is of no value whatsoever for long-term protection.

Babies possess a congenital immunity against infections such as measles; this is passive, derived while in the womb from the bloodstream of the mother, who has a permanent active immunity of her own dating from a childhood attack. Because the infant's resistance is borrowed, it lasts only a few months.

The production of antibodies which circulate in the blood is known as the *humoral* type of immune response. Another specific immune mechanism, the *cellular* response, is effected by small lymphocytes that recognise and respond to certain antigens, and migrate from the blood towards such antigens to protect the body against their effects. One or other response may predominate, but usually both are operant.

The reticuloendothelial system mediates both the early phagocytic response and the long-term immune reaction. It produces both antibody and lymphocytes in lymph nodes, tonsils and spleen, and in the reticulum of the bone marrow; in the latter the reticulum cells are also the precursors of red cells and polymorphs, i.e. the reticulum cell can develop into a lymphocyte, polymorph or red cell according to its situation.

The *lymphatic system* consists of lymphatic vessels and lymph glands, or nodes; the spleen is the largest mass of lymphoid tissue. The lymphatics form plexuses in all the body tissues except the central nervous system (see Fig. 5.13). They drain into trunks with one-way valves, allowing fluid to pass only towards the heart, accompanying the major blood vessels. In the limbs the nodes are situated mainly at the bends of elbow and knee, and the groin and armpit. Other nodes are sited close to the main viscera in the abdomen, in the mesentery, and along the aorta and inferior vena cava; others at the root of each lung and the bifurcation of the trachea have acquired them from polluted inspired air.

An individual node is pinkish-grey and bean-shaped (see Fig. 15.3). Afferent lymphatics drain into it and efferent vessels leave, carrying lymph with added lymphocytes and antibodies. New lymphocytes are constantly formed in the germinal centres, and enter the blood or lymph draining the node; they also re-enter the node from the circulation.

The lymph node is a filter to which afferent lymph brings debris or bacteria for phagocytosis by reticuloendothelial cells. It also receives antigen and reacts by a combined humoral and cellular response. It produces new lymphocytes and is a staging centre for old ones.

The *spleen* (see p. 164) can be regarded as a very large specialised lymph node. Its reticuloendothelial content carries on phagocytosis of effete red cells and other debris from the bloodstream. In certain conditions, such as severe anaemia, red cells are, abnormally, produced there. It can contract and expel blood into the circulation to meet emergency demands, as in exertion. It is also a source of lymphocytes and antibodies.

As the spleen removes from the circulation worn-out red and white cells and platelets, it contains a good deal of iron pigment. Whenever red cells are increasingly destroyed, the spleen enlarges to cope, e.g. in malaria, where the parasite inhabits the red cells. Splenic enlargement is usual in countries where malaria is endemic.

However, the spleen is not essential to life. It is often damaged in accidents and can be removed without harm, because its functions are taken over by other organs of the reticuloendothelial system.

The lymphocytes manufacture little or no antibody, but in response to antigenic stimulus they become able to 'recognise' antigen, and to crowd round it at local sites. In addition, the lymphocyte can transform itself into a larger, more primitive cell that does form antibody.

Lymphocytic immunity is important in the reaction of grafts from

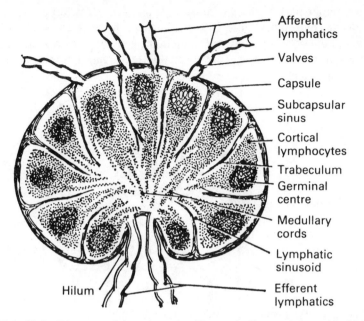

Fig. 15.3 Diagram of lymph node (From *A Companion to Medical Studies, Vol. 1*)

other individuals. Tissue material from any other person, except an identical twin, is recognised as foreign and cast off. A badly burned child may be skin-grafted from its parent with apparent success, but the grafts slough off after a few weeks, due to local attack by lymphocytes. It depends on the fact that an individual contains antigenic substances present in no other person. Similar considerations apply to the transplantation of whole organs, such as the heart or kidney.

All this relates to the *host versus graft* reaction. There is also a *graft versus host* reaction; certain cell transfusions may damage the liver, lungs, kidneys and lymphoid system, most frequently transfusion of blood or blood fractions.

Thymus
The thymus is a large lobed structure behind the upper part of the breast bone, covering the great vessels arising from the heart and

the upper part of the pericardium (Fig. 15.4). It is largest in the young child and shrinks in the adult to a mere remnant. It is a lymphoid organ but does not share in the general lymphatic circulation. Many lymphocytes are formed in the organ, but most never leave the gland.

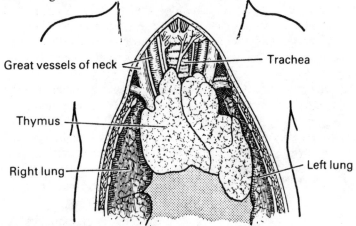

Fig. 15.4   The thymus in the new-born child (After *Gray*)

In the foetus and infant, the thymus seems to control the development of lymph nodes and lymphoid tissue. If it is absent, these fail to develop, lymphocytes are few and death results from infection because of a failure of immune processes. In the adult the thymus is a source of fresh lymphocytes for immunological imprinting, but it is not the only one and its removal causes little harm, although there may be some diminution of the capacity for cellular immune response. After heavy irradiation, the activities of bone marrow and lymphoid tissue are seriously depressed, sometimes temporarily abolished; the thymus is more resistant, and controls the restoration of lymphocyte production and cellular immunity. The importance of this is that irradiation is often used to suppress an unwanted immune response, e.g. after a transplant.

The thymus seems to be important in 'immunological surveillance', or the capacity to recognise certain antigens as foreign and to react against them. The converse of this is 'immunological tolerance'. An individual is tolerant of his own tissues, and does not normally react against them. He is also tolerant of foreign antigens encountered in the foetal or neo-natal period. In certain circumstances, the immune response can be stimulated by an individual's cell antigens and can attack his own tissues, resulting in various diseases such as thyroiditis or rheumatoid arthritis. The thymus is actively involved in this process, in which a self-antigen comes to be regarded as hostile. The protective antibody reaction may occur with such violence as to endanger life, when there is a long interval between the first and second dose of a sensitising agent. Then the neutralisation occurs not in the bloodstream but in the tissues, and severe shock and collapse – even death – may result. This reaction, known as *anaphylaxis*, is rare in man.

The defensive activities of the reticuloendothelial system may be defective from birth or become so in later life, so that we may speak of a *congenital* or *acquired immune deficiency syndrome*. Until recently, the usual cause of the acquired syndrome was the action of steroidal or other drugs deliberately used as *immunosuppressants* to prevent the rejection of grafted organs, such suppression not usually outlasting the treatment. But recent years have seen the spread of AIDS, an infection in which immunosuppression is brought about by a virus now usually known as HIV: human immunodeficiency virus. This virus is transmitted by sexual intercourse, or by inoculation, or via transfusion of blood or blood products, or by the mother to the child in the womb.

The disease has a long incubation period, during which the host is potentially infective to others. Viral DNA is permanently incorporated into the DNA of many host cells and when these foreign genes eventually dictate the formation of new viral particles the destruction of host cells begins. The virus is attracted to, and destroys, circulating lymphocytes of the immune system, but it also infects other types of reticuloendothelial cell in many organs and may directly attack the cells of the nervous system. Thus, the clinical manifestations include (*a*) 'opportunistic lung and other infections, as well as cancers, due to failure of immunologic surveillance, and (*b*) progressive dysfunction of the nervous system, lungs, bowel, skin and other organs. Within five years of infection about a quarter of the victims develop full-blown clinical disease; this is inevitably fatal and most die within three years of diagnosis. It may be that the ultimate mortality for all infected persons 20–30 years after infection is around 50%, i.e. that many of those infected do not develop the full clinical disease, but this is still far from certain.

16

The Heart and Circulation

The circulation is a closed circuit, round which the blood is propelled by the contractions of the heart. Blood is driven into the *arteries*, thickwalled elastic tubes which by their recoil aid its distribution. The arteries divide into smaller and smaller branches and finally into a meshwork of fine *capillaries* – microscopic thinwalled vessels that pervade every tissue except the cornea of the eye, the outer layer of the skin and articular cartilage. This meshwork joins up again to form small *veins*, which become larger trunks as they travel centrally towards the heart. The veins are thinwalled, have no pulse and contain valves to prevent any backward flow.

The smaller branches of arteries are called *arterioles*. These arterioles communicate freely by cross-branches called *anastomoses*, which ensure an adequate blood supply to any part, even when one vessel is obstructed. In a few situations there are no anastomoses and the local artery is an *end-artery*, the only blood supply; the retinal artery is an end-artery and its occlusion causes blindness.

The circulation is a device for transporting materials to maintain the internal equilibrium of body tissues and fluids. Its functions are to transport oxygen from lungs to body cells and carbon dioxide from cells to lungs; to transport nutrients to the cells and wastes from cells to kidneys; and to transport excess heat to the skin, to be lost to the exterior, so as to keep the body temperature constant. The blood is also a vehicle for a number of control systems, e.g. the feedback mechanism between blood glucose level and insulin pro-

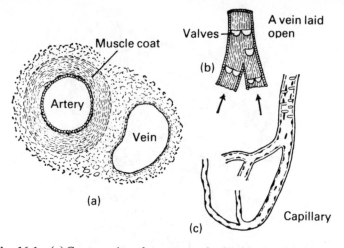

Fig. 16.1 (a) Cross-section of an artery and vein, showing the much thicker muscular coat of the former. (b) Vein slit open longitudinally, showing the valves with their pockets directed towards the heart. (c) A capillary with its delicate walls

duction, or between state of dilution and the pituitary activity (p. 300).

There are two separate circulations: a *systemic* circulation, concerned with the body as a whole and driven by the left side of the heart; and a *pulmonary* circulation, concerned with passage of blood through the lungs and driven by the right side of the heart. The two sides of the heart are separate, each with an upper chamber or *atrium* receiving blood from the great veins and a lower chamber or *ventricle* discharging blood into the great arteries. Stale venous blood from the body enters the right atrium, passes to the right ventricle, and is expelled through the pulmonary artery to traverse the capillaries of the lungs. Here it receives fresh oxygen from the air sacs, and gives up carbon dioxide to be exhaled. The fresh blood returns from the lungs in the pulmonary veins to the left atrium, then down to the left ventricle, and is distributed by the great artery of the body, the aorta, to the head, trunk and limbs. In the tissues the blood becomes dark and venous, and is collected up into two great veins: the *superior vena cava* draining the head and arms, and the *inferior vena cava* draining trunk and legs.

Venule

Arteriole

Fig. 16.2 Diagram of a capillary network and its supplying and draining
  vessels (From *A Companion to Medical Studies, Vol. 1*)

The arteries of the body contain bright blood and the veins dark
blood; the reverse is the case for the pulmonary vessels because the
lungs are concerned with reversing these chemical states.

There is a special arrangement of the abdominal vessels. Whereas
the veins leaving most structures pass directly to the heart, those
from the stomach and intestines enter the liver, where they break up
into a second set of capillaries so that the blood is filtered through
the liver before reaching the heart. This is to ensure that the liver
utilises and stores foodstuffs absorbed from the bowel, and is known
as the *portal system*.

Physiological features of the general circulation

The circulation is a closed system and, although the calibre of the
blood vessels varies enormously, the blood never escapes into free

contact with the tissues. The heart acts as a boosting mechanism in a pipeline, for it receives blood from the veins at low pressure and pumps it to the arteries at a high one. When the blood eventually reaches the thin-walled capillaries, the plasma diffuses through to mix with the tissue juices, which bathe the cells.

There is a pressure gradient from the high level in the great arteries to an intermediate point in the capillaries and still lower in the veins, which may be zero; circulation in the veins is largely dependent on outside assistance: the sucking action of respiration, which draws blood into the chest, the contraction of the muscles around the veins in the limbs. Thus, on walking, the leg veins are emptied by the action of the muscles, and their valves prevent reflux and direct the flow to the heart.

This difference between veins and arteries is reflected in structure; the veins are thin-walled with little muscular or elastic tissue; the arteries have thick coats with much muscular and elastic tissue. The arterial tree is an elastic distributing system with a recoil to the heartbeat, which is transmitted to produce the pulse felt at the wrist or temple. The contractile power of the arteries maintains a higher blood-pressure at a greater distance from the heart than would a system of inelastic tubes. The smaller arteries and arterioles can also exert a selective action on the circulation in particular regions by constricting or dilating to shut off or increase the blood flow. The veins form an inelastic reservoir which may become a stagnant pool if their valves become inefficient or the heart fails.

The flow in the veins is slower than in the arteries because of their greater total cross-sectional area, though the volume of blood transmitted per minute must be the same. In the arteries, there is friction between the blood and the vessel wall, and the blood cells tend to cling to the periphery so that flow is most rapid in the centre.

As the arteries divide, the cross-sectional area of their branches increases until the total sectional area or 'capillary bed' may be 1000 times that of the aorta. This is another aspect of the progressive fall in blood pressure as the arterioles branch into capillaries; when all the capillaries are widely dilated, as in shock, the blood-pressure falls very seriously.

The heart

The heart is a pump which continues to beat regularly and continuously for seventy years, with a rest interval never longer than a fraction of a second. This rhythmic contraction is inherent and independent of nervous control; chicken hearts have been kept contracting in culture media for many years.

The heart beats about seventy times a minute at rest, more rapidly in children and up to a hundred and fifty times a minute in the child in the womb. Contraction is known as *systole*, relaxation as *diastole*. Each beat begins as a simultaneous contraction of both atria expelling blood into the ventricles; this is followed by ventricular contraction, which throws the blood into the great arteries. This cycle lasts eight-tenths of a second, of which atrial contraction makes up one-tenth, ventricular contraction three-tenths and diastole four-tenths. During diastole the atria act as passive receptacles draining the great veins; since they contract a little before the ventricles, while the latter are still relaxed, the atria do not have much work to do and are relatively thin walled. The ventricles have the resistance of the peripheral circulation to overcome, notably the elasticity of the arteries, and are thick-walled and powerful. The right ventricle is only concerned with driving the blood through the

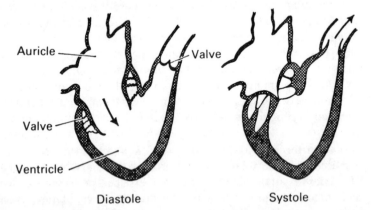

Diastole   Systole

Fig. 16.3   In relaxation or *diastole* the blood flows into the ventricle from the auricle through the open valve between. In contraction or *systole* the valve is slammed shut as the ventricle expels its blood into aorta or pulmonary artery

lungs; it has less work than the left, and its wall is only a quarter as thick.

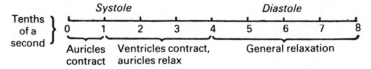

The regular sequence of the cardiac cycle is due to two little foci of sensitive connective tissue which trigger each contraction. These *nodal points* are stationed one at the junction of great veins and atria, one at the atrioventricular junction. They are connected to the heart muscle by a cable of fibres which runs down the septum between the two halves of the organ, giving a branch to either side. The heart is flaccid in diastole; when it contracts it becomes thick and cone-shaped, and transmits an impulse to the front of the chest on the left side. This is the *apex beat*, between the fifth and sixth ribs. If one listens with a stethoscope or the unaided ear, over this point, two characteristic sounds can be heard with each beat, coming closely together and repeated after the diastolic pause. These may be represented as follows:

. . . lubb, dupp . . . lubb, dupp . . . lubb, dupp . . .

and are both associated with ventricular contraction. The first is the slamming shut of the parachute-like valves that prevent reflux from ventricles into atria. The second occurs at the end of ventricular contraction, when the valves at the mouths of the great arteries fall together to prevent backflow into the ventricles.

Contraction of the heart chambers must be coordinated. An uncoordinated contraction is known as *fibrillation*, which tends to prevent any proper pumping action. Atrial fibrillation is common; it is not serious and is controllable by the drug digitalis. Ventricular fibrillation is rapidly fatal if not treated promptly.

Although the two halves of the heart constitute a single organ, they perform separate functions. The atrium and ventricle on each side constitute a pair of chambers concerned with pumping blood through a particular territory; and, although the blood traverses each pump in turn, it is possible to imagine the right and left halves as separate organs fixed together for convenience (Fig. 16.4). The output of the heart must obviously be the same on both sides, or

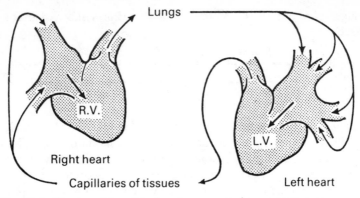

Fig. 16.4   The two sides of the heart represented as separate organs

blood would be dammed up on one or other side. This output, at rest in the adult, is some 3-4 litres a minute, i.e. the whole of the blood in the body is passed through the heart in under two minutes.

The *stroke volume* is the amount of blood ejected in systole from the left ventricle into the aorta, some 60–100 ml/min. The output per minute is equal to this quantity multiplied by the pulse rate.

The control of the heart

The heart's activity is varied by altering either the *heart rate* or its power of contraction, i.e. its *output*. The factor most influencing output is the inflow from the great veins, for output must equal this, whatever the resistance of the general circulation, or blood will be dammed up in the venous system; this happens in heart failure. Venous inflow is greatly influenced by the sucking action of respiration. Breathing in produces a negative pressure within the chest which draws blood into the great veins and right atrium from the venous trunks of the head, abdomen and limbs. Therefore, filling of the heart is largely due to ordinary breathing; a deep breath increases intake, the output rises correspondingly and the pulse quickens at the wrist. Conversely, if pressure within the chest is raised by straining, the great veins are flattened out, cardiac output falls and the pulse weakens.

Mechanically, the work of the heart, i.e. the energy produced in ventricular contraction, is used up in two ways. First, blood expelled acquires a definite velocity and momentum, and, secondly, the resistance of the arterial tree has to be overcome, the stretching of the arterial walls and the tendency to recoil providing a store of potential energy like a compressed spring. Most of the work done is in overcoming arterial resistance. During the course of a normal life, the work done by the heart is equivalent to lifting a weight of 10 tonnes through a height of 16 km.

The circulatory resistance is equal to the arterial blood-pressure, and the heart adapts to expel exactly its intake of blood whatever the arterial pressure, or to get rid of a varying intake against a constant resistance. This adaptability lies in the power of heart muscle fibres to contract more forcibly the more they are stretched, and vice versa. Obviously there are limits to the heart's ability to meet increased demand by dilating, with increased contraction and hypertrophy of its fibres. One limit is that its fibrous bag, or pericardium, cannot stretch; the other is the mechanical disadvantage at which stretched fibres have to work. This property of overcoming handicaps, such as a leaking valve or a raised blood-pressure, is known as *compensation*, and results in an organ larger and thicker than normal. But when the limits of adjustment are surpassed, *decompensation* results; output cannot keep pace with influx, and the veins become stagnant reservoirs.

The blood-pressure within the heart can be measured by inserting a cardiac catheter, a tube passed from an artery or vein of the arm or leg into the organ. The pressure in the systemic circulation is five times that on the pulmonary side. The stabilisation of blood-pressure is important, particularly for the kidneys and brain. The kidneys stop secreting urine if the systolic pressure falls below 50 mm of mercury; and one cannot survive more than a few minutes if the brain is deprived of arterial blood.

The nervous control of the heart

Apart from its self-regulating mechanism, there is a nervous control to adapt the organ to the needs of the organism. This is effected by branches of the two constituents of the independent autonomic nervous system – the sympathetic and parasympathetic. As with the bowels and bladder, the two are mutually antagonistic but work together to achieve a balanced control (Fig. 16.5).

The *parasympathetic* fibres are derived from a nerve called the *vagus*, running from the base of the brain through the neck and

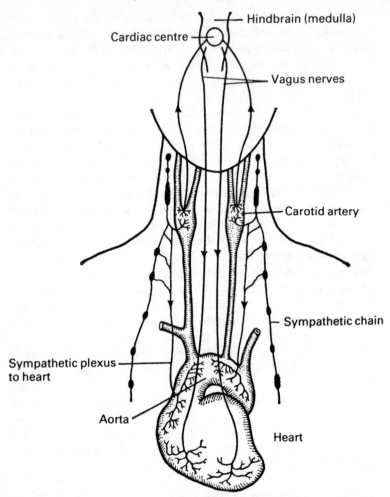

Fig. 16.5 The nervous control of the heart; anatomical arrangements
(Adapted from Starling's *Principles of Human Physiology*)

chest. They depress heart action, slowing its rate and weakening its force. If the nerve is stimulated experimentally, the heart may stop; firm pressure on the eyeball excites the vagus and slows the pulse. Vagus action is achieved by the liberation at its endings in the heart muscle of *acetylcholine*.

The *sympathetic* fibres are derived from the sympathetic chain, which runs each side of the spinal column. They accelerate the heartbeat and increase the force of contraction, by liberating *adrenaline*. Adrenaline from the adrenals also reaches the heart via the circulation at times of stress, producing a rapid, forceful beat.

The ultimate cell-stations from which both sympathetic and parasympathetic fibres derive are in the hindbrain, or medulla, where they form two centres, one inhibiting and the other augmenting cardiac function. These exert central control on the basis of information received from various parts of the body, in particular from stretch or pressure receptors in the heart and great arteries. These messages indicate the filling of the heart, the blood-pressure, and the extent of cardiac dilatation; the control centres react to secure that the heart beats faster and more forcefully if venous influx increases or the arterial pressure falls, and slows and beats less strongly if these changes are reversed.

The heart gets its own blood supply from two tiny vessels, the

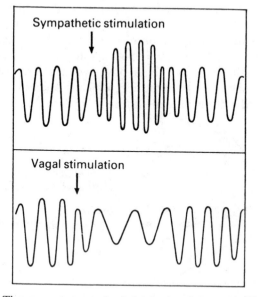

Fig. 16.6 The nervous control of the heart. Sympathetic stimulation produces a rapid forceful beat, parasympathetic action a slower beat of less amplitude

coronary arteries, which spring from the aorta at its commencement and ramify over the surface of the organ. These are very sensitive to nervous control and changes in oxygenation of the blood, and increase the blood supply to the wall of the heart when extra demands are made on it. Coronary disease (atherosclerosis) or obstruction by clot (coronary thrombosis) interfere with the heart's capacity to respond to demand. The great energy output of the heart is obtained not so much from glucose – as with skeletal muscle – as from free fatty acids, keto-acids and lactate in the blood, and the heart muscle is sensitive to changes in concentration of sodium, potassium and calcium ions.

The circulation

There is a *pressure gradient* in the circulation, falling from the left ventricle through the arteries and capillaries to very low or zero in the veins. This pressure is measured in terms of a column of mercury. In all the arteries it is about 120 mm of mercury but it falls to 20 mm in the capillaries and to only a few millimetres in the main veins. In the large veins of the neck and chest the pressure is often negative because of the sucking action of respiration. When cut, arteries bleed from the end nearer the heart, in forcible spurts. Veins bleed from the farthest end in a gentle flow. Wounds of the veins of the neck may suck in air from outside; which is dangerous for the air gets churned in the heart and obstructs the blood flow.

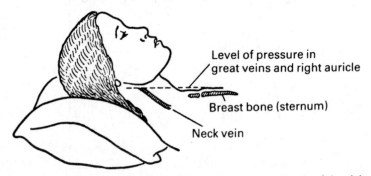

Fig. 16.7   The basic level of venous pressure in the reservoir of the right atrium of the heart is shown by the extent of the visible column of blood in the superficial veins of the neck

The blood pressure in the arteries is alternately high and low, corresponding to contraction and relaxation of the left ventricle. The *systolic pressure* is some 130 mm of mercury, the *diastolic pressure* only 70 mm, and the fluctuation between is the *pulse pressure*, i.e. 60 mm. In the veins and capillaries the pressure is steady. The term 'blood-pressure' in common usage refers to arterial pressure and is expressed in double figures as 130/70, though there is a considerable range of variation. Since the higher systolic figure depends on contraction of the left ventricle, and this is influenced by emotion and exertion, it is the steady diastolic pressure that provides a base line.

The intermittent flow from the heart into the arteries becomes a constant flow when the capillaries are reached, due to the recoil of the arteries and arterioles, which continues the driving force while the heart itself is relaxed.

The time taken for the blood to pass from any given point through the heart and lungs and back to the same point is known as the *circulation time*, and is about 25 seconds at rest. This does not imply that the speed of the blood is constant, as this varies with the heartbeat and the size of the blood vessel; at its maximum, in the arteries springing from the heart, the velocity is about 1·6 km/h. Nor must we confuse the velocity of blood flow with that of the pulse wave; this is a pressure wave transmitted at some 24 km/h through the blood in the arteries and is unrelated to any forward movement of the fluid itself.

The network of tiny arterioles and capillaries, or capillary bed, provides the main resistance to the thrust of the left ventricle. But it is not a fixed resistance, for the arterioles have muscular coats and the capillaries contractile cells, which are subject to nervous and chemical control. We speak of *vasoconstriction* when these are narrowed and *vasodilatation* when they open up. This is a means for adjusting the dynamics of the circulation in different parts, an arrangement necessary to meet the changing effects of gravity in different postures.

In standing, the whole of the blood would collect in the small vessels of the abdomen and legs were these not largely shut down by vasoconstriction. Such a widespread narrowing causes a rise of blood-pressure throughout the body, as the remaining vessels are overstretched; conversely, extensive vasodilatation leads to stagnation and a fall of blood-pressure. The enormous vascular area of the abdominal organs forms what is known as the *splanchnic pool*, in contrast to the *somatic* capillary bed in the skin, muscles and skeleton. When the splanchnic vessels are dilated, as after a heavy meal, the blood-pressure falls and the blood supply to the brain is decreased. We become sleepy and tend to lie down because of a temporary cerebral anaemia and the circulation can be maintained more easily in the horizontal position. Fainting is an acute change of the same nature after a shock which produces splanchnic vasodilatation.

Often, in old people, this control mechanism is inadequate and there is a sudden drop in blood-pressure, or postural hypotension, on standing, which may cause falls or fainting attacks. In *surgical shock*, which develops after severe injuries or burns, there may be such widespread expansion of the capillaries that the body 'bleeds to death into its own blood vessels'; the blood-pressure falls profoundly and too little blood is available for the heart to work on or the lungs to aërate.

The addition of fluid to the blood has little effect if it is crystalloid; the excess water and salts rapidly pass into the tissue spaces. If the fluid has a colloid content, this cannot happen and the vessels dilate to accommodate the extra fluid without causing any rise in pressure; meanwhile, the kidneys are stimulated to pass more urine. The loss of blood, as in donation of, say, 600 ml, is compensated so efficiently that the circulatory effects are negligible. But a sudden loss of several pints causes a severe fall in blood-pressure which must be compensated. There is general vasoconstriction, the heart beats faster, sweating occurs, the skin becomes cold, kidney blood flow is reduced and urinary output falls. At the same time, fluid enters the circulation from the tissue spaces. Thus blood-pressure and blood volume are restored, though the blood is more dilute, i.e. it contains less red cells and haemoglobin.

The circulation is like a railway system with numerous sidings but a constant number of trains. If some of these are shunted off, the main line traffic will be less heavy, i.e. the blood-pressure will fall. If all the trains are in use, the traffic is intense, i.e. the blood-pressure rises. Such a regulation occurs in all parts; any inflammation sets up a local vasodilatation which flushes the part with blood. The arterioles may be so widely opened that they communicate their pulse to the normally pulseless capillaries and the part throbs.

The mechanism controlling the small vessels is like that of the heart – part nervous, part chemical. The sympathetic fibres that accelerate the heart constrict the vessels and prepare the body for emergency for they divert the blood from the skin and non-essential organs to the muscles and the brain.

The parasympathetic fibres promote relaxation and fall of blood-pressure. However, the two systems are reciprocal, and the vessels are never completely shut down and never completely dilated (except in severe shock).

The arterioles and capillaries are very sensitive to circulating adrenaline, which has the same effect as the sympathetic, and are dilated by substances liberated from damaged tissues, e.g. *histamine*. As in the heart, the ultimate controlling centres are in the medulla. The prime function of these centres is to ensure that the supply of blood to the brain is safeguarded; other tissues can survive deprivation of oxygen for perhaps an hour or more, but the brain cells die after more than a few minutes of anoxaemia.

This is seen after head injury causing progressive bleeding within the skull; this compresses the brain and makes it more difficult for blood to be pumped in from the heart. The vasoconstrictor centre in the medulla reacts by so shutting down vessels all over the body that the blood-pressure may rise to enormous heights; this forces blood into the brain-box, until the process breaks down.

Exercise

Although the heart and vessels adjust to a particular situation, they are anticipated by mental factors, so the emotion in waiting to start a race produces a rise in blood-pressure and pulse rate long before these are physically necessary.

During strenuous exercise, the oxygen demands of the muscles may be ten times those of the resting stage, and this must be met by an increased cardiac output. Arterial blood is already saturated with oxygen and cannot dissolve more; the only way to bring more oxygen to the tissues is to rush the blood round the circulation at a faster rate.

Bodily changes in exertion are complex. Muscular contraction speeds return of blood to the heart by milking it upwards in the veins; and this increase in venous inflow implies an equivalent increase in cardiac output if the heart is not to fail. The amount of blood in the circulation cannot be increased, but more can be diverted to the muscles by shutting down the vessels of the skin and the splanchnic area, and by contraction of the spleen. Exaggerated respiratory movements suck more blood into the great veins of the chest. The sympathetic system, aided by the pouring of adrenaline into the blood, increases the heart rate and the force of each beat, and mobilises liver glycogen as sugar for the muscles. The waste products of muscular contraction – carbon dioxide and lactic acid – stimulate the brain centres controlling the heart, the vessels, and the rapidity and depth of respiration.

17

Respiration

The respiratory system

Inhaled air enters the pharynx and travels through the air passage proper. The first part is the *larynx*, which leads on to the windpipe (*trachea*), and divides in the upper chest into a right and left *bronchus* for each lung. Each bronchus subdivides in the lung to form numerous *bronchioles* which end in clusters of tiny *air sacs* or *alveoli*; it is in the walls that interchange occurs between the gases dissolved in the blood and those of the inhaled air (see Fig. 5.10).

The chest cavity is divided into right and left by a partition in which the heart is embedded; the two halves are separate and contain the lungs. Each cavity is lined by a smooth membrane, the *pleura*, and each pleural space is a closed sac, for the membrane is reflected from the chest wall to cover the surface of the lung. Normally, there is no pleural cavity, for the two layers are in contact, each lung filling its side of the chest. But the lung is very elastic, tending to shrink and expel its contained air; since in health it cannot do this, there is a negative pressure in the potential pleural space. The moment air is admitted, the lung collapses into a small, solid, airless mass.

Respiration

The purpose of respiration is the entry and exit of air to and from the lungs. In *inspiration* the chest cavity is enlarged and air enters; in *expiration* the reverse occurs. Respiratory movements may be thoracic or abdominal, or both. In thoracic inspiration the breast bone is lifted, and the ribs are elevated and lie more horizontally; the chest is increased in diameter from side to side and from front to back. In abdominal inspiration the diaphragm contracts down on the abdominal organs, bulging the abdominal wall and increasing the height of the chest. Inspiration is active, due to muscular exertion, whereas expiration is passive; the chest wall subsides, the abdominal muscles recoil, the diaphragm relaxes, and air is driven out of the lungs.

Lung ventilation

During quiet breathing, about 600 cm^3 of air are taken in and breathed out at each respiration; this is the *tidal air*. By deep inspiration, some 2000 cm^3 more can be taken in, the *complemental air*; and after a normal expiration it is possible forcibly to expel a further 1200 cm^3, the *supplemental air*. The sum of these figures, the total possible flow in one direction, is known as the *vital capacity* and represents the maximum volume of air that can be expelled after as deep an inspiration as possible, i.e.:

Complemental air	= 2000 cm^3
Tidal air	= 600 cm^3
Supplemental air	= 1200 cm^3
Vital capacity	= 3800 cm^3

The vital capacity varies a good deal, depending on physique and fitness; it is greatest when standing and least when lying down, and is diminished when the ribs are fixed by arthritic disease. But no expiration, however forcible, can empty the lungs of air; a minimum is left in the bronchi and air sacs, the *residual air*, some 1500 cm^3. Finally, not all the 600 cm^3 of tidal air will reach the air sacs, for some is needed to fill the air passages themselves, a *dead space* of some 150 cm^3.

The air passages are not rigid tubes. They have a strong muscular coat which can alter their calibre by constriction or relaxation, so varying the volume of the dead space and the resistance to the flow of air. Asthma is an intense spasm of bronchial muscle and makes expiration a long-drawn-out effort; inspiration is less hindered. The smooth muscle of the bronchi is under the control of the parasympathetic and sympathetic nerves; the latter, in exertion and fright, open the air passages widely to achieve maximum possible influx.

Consequently, adrenaline, which stimulates sympathetic nerve endings, may be injected to combat asthma.

Chemical aspects of respiration

In the tissues some 400 cm^3 of oxygen is used every minute to burn food material, with the production of carbon dioxide. The oxygen is taken from the blood in the capillaries and the carbon dioxide returned in exchange; so the venous blood leaving any part contains less oxygen and more carbon dioxide than the arterial blood that entered it. The lungs reverse this state in the venous blood, restoring its oxygen quota and expelling excess carbon dioxide by interchange with the air in the air sacs. The venous blood is rendered arterial in the lung capillaries, and the changes in the tissues are reflected in changes in the composition of the tidal air, which, when breathed out, contains less oxygen and more carbon dioxide than on inspiration:

	Oxygen	*Carbon dioxide*	*Nitrogen*
Inspired air	20·95%	0·05%	79%
Expired air	16·5%	4·0%	79·5%

The content of nitrogen appears slightly greater in the expired air because slightly less air is breathed out than in. The reason is that the volume of carbon dioxide exhaled is slightly less than that of the oxygen for which it is exchanged. This ratio *or respiratory quotient*:

$$\frac{\text{carbon dioxide output}}{\text{oxygen consumption}}$$

is always less than unity and usually about 0·85 and depends on the nature of the foodstuffs being burnt (see p. 222).

The respiratory gases are carried dissolved in the blood, and the total gas that can be expelled is about 70%. Of this, the relatively insoluble nitrogen forms only a small proportion. Analysis of venous and arterial blood reveals their different gaseous content in volumes per cent:

	Oxygen	*Carbon dioxide*	*Nitrogen*
Arterial blood	19	50	1
Venous blood	10·5	58	1

The problem is not simply relative solubilities for it is affected by the following factors:

1   The amount that can be dissolved is proportional to the pressure of the particular gas in the lungs or tissues.
2   The gases are not held in simple solution but in loose chemical combination – oxygen with haemoglobin, carbon dioxide with the carbonic acid and sodium bicarbonate of the plasma.
3   The two gases tend to displace each other from their combinations with corpuscles or plasma.

Oxygen
When venous blood, with its reduced content of oxygen at low pressure, reaches the lungs, it is exposed to a higher pressure of oxygen in the alveolar air. The pressure gradient is towards the blood, and oxygen combines loosely with haemoglobin to form oxyhaemoglobin. When the blood has been swept through the heart to the capillaries in the tissues, it arrives at a region where oxygen has been used in combustion and is at a lower pressure than in the blood; the gradient sets the other way, and oxygen is set free from oxyhaemoglobin and diffuses into the tissues. This dissociation is catalysed by the presence in the tissues of the excess carbon dioxide formed by combustion. The very waste products formed by using-up oxygen stimulate the liberation of oxygen from its bound state in the blood.

Carbon dioxide
The pressure of this gas as a waste product in the tissues is much higher than in the fresh blood brought by the arteries. The gradient favours passage of the gas from the tissue fluid into the capillaries, where it enters into loose combination as carbonic acid and sodium bicarbonate in the plasma. This is carried with the venous blood through the right heart to the capillaries of the lungs, where it is exposed to freshly inhaled air with a very low carbon dioxide content; the gradient sets the other way and the gas is expelled with the exhaled air. This transference is assisted by the saturation with oxygen occurring in the lungs, as a high oxygen content in blood tends to break up loose carbonic compounds.

Tissue respiration

We have been describing 'external respiration', concerned only with gaseous interchange in the lungs and tissue capillaries and not with the vital processes in the tissue cells. This less obvious process is called 'internal respiration', and is the motive of all the complex apparatus of heart, lungs and circulating blood. Every tissue uses oxygen and liberates carbon dioxide for its energy requirements, in amounts depending on functional activity, greatest in the active tissues of the heart, brain, liver and muscles. Little is known about the fundamental processes in the cell between the intake of oxygen and the formation of carbon dioxide. There is a complex system of enzymes known as *oxidases*. Respiration, involving the use of oxygen, is known as *aërobic*. Many bacteria and other organisms are capable of respiration without using oxygen, though still forming carbon dioxide, an *anaërobic* process.

The regulation of respiration

Respiration is complex, involving the nostrils, larynx, chest wall, diaphragm and abdomen; accurate coordination is necessary. This is achieved by a *respiratory centre* in the medulla, one of the group of vital centres that includes the cardiac and vasomotor centres. The centre is constantly influenced by information from the lungs and other parts, and sensitive to changes in the oxygen and carbon dioxide content of the blood in the cerebral vessels, so that it may alter respiration as needed. It is automatic and unconscious, though we can control breathing with an effort. It is one of the most fundamental activities; in progressively deep anaesthesia or poisoning, respiration is one of the last bodily functions to stop.

The *chemical control* of respiration is effected by response of the centre to changes in the oxygen and carbon dioxide content of the blood; the carbon dioxide is more important, for it is essential to get rid of excess waste products, whereas the oxygen content of the air is ample for respiration within wide limits. The slightest rise of carbon dioxide increases respiration, and an increase in its content in the air to only 1·5% multiplies ventilation by half. In progressive asphyxia, with accumulation of carbon dioxide, respiratory convulsions develop, followed by collapse and death. There is much less sensitivity to oxygen lack, and the atmospheric level can fall from 21% to 13% without noticeable effects; even then there is no great distress and collapse follows without warning – as in unmasked pilots at high altitudes.

The *nervous* or *reflex control* is effected by the vagus nerves, which connect the vital centres of the medulla with the heart, lungs and abdominal organs. On the sensory side, these provide the respiratory centre with information as to distension of the lungs, in blood-pressure, and gas content in the great arteries. On the motor side, they carry reflex responses, modifying the action of the respiratory muscles.

Of the two parts of the respiratory centre, the inspiratory one is dominant. Inspiration is active, due to messages sent along the vagus; expiration is a passive relaxation. The coordination of the two is a pretty example of reflex action. When inspiration is at its height, the lungs are inflated and their sensory nerve endings stretched; they stimulate the expiratory part of the nervous centre, vagus action is inhibited and expiration follows. The very act of inspiration produces the stimulus for the succeeding expiration.

The nervous control of respiration affects its rhythm, the chemical control its depth, and these reciprocate; if increased carbon dioxide increases the volume of inspiration it must also accelerate the stretch-response and thus the nervous control of rhythm. In exercise the rising carbon dioxide in the blood as a waste product activity reaches the brain and at first stimulates respiration. But, as effort increases, carbon dioxide is washed out of the blood by forced breathing, and the stimulus now comes from the lactic acid formed in contraction and from secreted adrenaline. It is this steady stage that is one's 'second wind'. Deliberate overbreathing at rest will so empty the blood of carbon dioxide that the normal stimulus to respiration is removed; there is no wish to breathe during the next couple of minutes, but the experiment is a little dangerous.

Oxygen deficiency

A lack of available oxygen may be due to a lowered oxygen content of the air; obstruction to the air passages; lung disease; inability of the blood to carry its full quota of oxygen, as in haemorrhage or anaemia, or when the haemoglobin is put out of action by a poison such as *carbon monoxide*; when the circulation is slowed in heart failure; and inability of poisoned tissues to use oxygen. All these result in overbreathing at rest, in an effort to make up for the deficiency by increasing the intake of air; the cardiac invalid may pant as strenuously as a man who has run a mile. In all these states the blueness of insufficiently oxidised haemoglobin is evident in the lips and under the nails – the condition of *cyanosis*.

Heights and depths
Similar changes occur at altitudes; the oxygen available decreases as atmospheric pressure falls and reaches the danger point of 11% at 4877 m, above which an oxygen mask is desirable. Immediate compensation is by overbreathing and by an increase in the rapidity of the circulation. More permanent adaptation can be made by mountain dwellers, who acclimatise by an increase in the number of red corpuscles and an habituation of the tissues to a lowered oxygen content.

In working at depths or in compressed air, the pressure of carbon dioxide is increased proportionately and produces its usual toxic effects. At the same time, nitrogen, normally poorly soluble in blood, is driven into solution; and, if the pressure is suddenly released, is liberated in the blood as bubbles which may block the heart or nervous system and cause death or paralysis, hence the necessity for gradual decompression.

18

Excretion

The gross structure of the urinary tract is set out on pages 76–7. Microscopically, each kidney contains over a million secretory units, or *nephrons*, each nephron comprising a tuft of blood capillaries, or glomerulus, invaginated into a capsule leading to a convoluted tubular system which discharges with others into a larger collecting tubule, draining at the apex of a pyramid into the renal pelvis (see Fig. 8.17). Such is the functional reserve, that two-thirds of one kidney will carry out the excretory function essential to life.

Each tubular system is closely enveloped by a plexus of blood vessels derived from a branch of the renal artery. From this the blood is gathered into a branch of the renal vein, which removes the blood after its purification.

The general function of the kidney is to maintain a constant internal environment. It is a very efficient mechanism for preserving constant chemical and physical composition of the blood plasma and the extracellular fluid. The kidneys regulate water content, pH and osmotic pressure so as to maintain electrolyte and water equilibrium. The blood and tissue levels of mineral ions, organic metabolites and wastes must be kept between narrow limits to preserve health. Removal of the kidneys causes a piling-up of urinary constituents in the blood (uraemia), with death after two to three weeks.

Practically all the waste products of metabolism are excreted in the urine. The only exceptions are (*a*) carbon dioxide, exhaled from the lungs, and (*b*) breakdown products of bile and other digestive juices, expelled in the faeces. If the blood contains an abnormal non-colloidal constituent, or an excess of normal constituents such as water and salts, the kidneys excrete these to restore normality. The kidneys are the only means for eliminating the end-products of protein metabolism. But they do not themselves form any of the wastes they excrete – they merely rid the blood of substances formed in other parts of the body. In renal failure this process is inefficient, the blood levels of urea and other wastes rise enormously, and the tissues become poisoned by these metabolic products.

To fulfil these functions the kidneys possess a huge blood supply; they are a pair of filters through which approximately 1 litre of blood circulates each minute, i.e. the whole of the blood in the body passes through them in 5 or 6 minutes, or 1800 litres a day, which is 400 times the blood volume.

Urine

The water content of the urine is derived partly from ingested food and fluids, and partly from oxidation processes in the tissues. So its exact composition will vary with the food, the fluid intake and fluid loss by other channels. Nevertheless, it remains fairly constant.

It is a clear yellow colour – paler when diluted by excessive drinking, darker when concentrated by sweating. It contains mucus, derived from the lining of the urinary channels. It is usually moderately acid. The total output in twenty-four hours varies with fluid intake and outside temperature – it averages 1·5 litres, with a range of 1–2·5 litres – and its formation is lowest in sleep and maximal during the day or on exertion. Its specific gravity varies around 1·01–1·03.

The urine may be turbid from calcium phosphate. It may deposit mucus, urates, crystals of calcium oxalate or phosphates. Microscopically, deposits include occasional casts shed from the lining of the renal tubules, sometimes red cells after exertion. Urine is frothy when shaken due to the presence of bile salts.

Its constituents are organic and inorganic. The *inorganic* group includes chlorides, phosphates, ammonia, sulphates, magnesium, calcium and iron. The ammonia is derived from protein breakdown, sulphates result from oxidation of sulphur in protein; phosphates partly from the food and partly from oxidation of organic phosphates in the tissues. The *organic* components are mainly nitrogenous end-products of protein metabolism – urea, uric acid, creatinine, and so on. Urea is the most important, and the urea content is an important index of protein metabolism. There is an excess of uric acid in *gout*; this is a breakdown product of nucleoprotein. There are also various organic *pigments*, derived from bile pigments. The non-nitrogenous organic constituents include oxalic acid from the food (e.g. rhubarb), lactic acid (after exercise), and derivatives of water-soluble hormones and vitamins.

Certain *abnormal* constituents appear in the urine at times. If the kidneys are diseased, some of the normal serum proteins enter the urine, which coagulates on warming. Normally, the urine only contains very little protein and only under certain conditions: after exertion, in pregnancy and after long standing in certain people. Glucose is found in diabetes, lactose in nursing mothers and ketone bodies in severe diabetes or starvation. The urine may contain haemoglobin, after an incompatible blood transfusion or in malignant malaria, or increased bile salts and pigments in liver disease and jaundice.

Filtration

The formation of urine entails the passage of water and crystalloids from the blood capillaries of the glomerulus into the nephron capsule and tubules. Colloids, such as the proteins of blood, do not normally appear in the urine. The crystalloids are inorganic mineral salts, and organic compounds like urea and uric acid. The essence of renal activity is that these substances are *concentrated* in the urine. The water content of urine and blood is much the same, but there is sixty times as much urea in urine as in blood, fifteen times as much uric acid, forty times as much phosphate, seven times the amount of potassium, twice as much chloride and so on.

The basic processes are those of *filtration* in the glomeruli, and *reabsorption* of water and certain substances in the tubules. The energy required for filtration is considerable, for the osmotic pressure of the plasma proteins tends to hold the excreted substances in the circulation. The driving force that overcomes this is the blood-pressure in the glomerular arterioles ultimately derived from the pumping action of the heart. Should this fall below the osmotic pressure of the blood proteins (40–50 mm of mercury), as in severe shock, urine formation ceases. Thus urine formation begins as

Table 18.1 Relative composition of plasma and urine in normal men

	Plasma g/100 ml	*Urine* g/100 ml	*Concentration* *factor*
Water	90–93	95	—
Proteins and other colloids	7–8·5	—	—
Urea	0·03	2	×60
Uric acid	0·002	0·03	×15
Glucose	0·1	—	—
Creatinine	0·001	0·1	×100
Sodium	0·32	0·6	×2
Potassium	0·02	0·15	×7
Calcium	0·01	0·015	×1·5
Magnesium	0·0025	0·01	×4
Chloride	0·37	0·6	×2
Phosphate	0·003	0·12	×40
Sulphate	0·003	0·18	×60
Ammonia	0·0001	0·05	×500

ultrafiltration of a large volume of blood plasma from the glomerular capillaries into the capsular space, colloids such as proteins being held back while crystalloids pass through. The only difference between plasma and this initial filtrate is the absence from the latter of molecules above a certain borderline size. The filtrate contains all the substances present in the blood, and in the same concentration, except the colloids and fats.

Tubular absorption

Urinary secretion is not, however, a matter of simple filtration, with reabsorption of water back into the blood in the tubules. There is also a *selective* action in the tubule cells which accounts for the fact that, while there is sixty times as much urea in the urine as in the blood, there is only twice as much chloride. The filtrate from the glomeruli enters the tubules, where most of the water and some electrolytes are reabsorbed into the bloodstream, while wastes such as urea are selectively retained in the urine.

80–85% of the water and sodium in the glomerular filtrate are reabsorbed in the tubules. There is, for each constituent of the blood, a threshold concentration which must be exceeded before it appears in the urine. Thus, sugar is normally never excreted because its threshold is rarely exceeded, save in diabetes; in other words, all the glucose arriving in the filtrate is reabsorbed. On the other hand, reabsorption of chlorides is much less, and so chlorides are normal constituents of urine. Further, the urine may be acid or alkaline and so compensate for excessive intake of alkalis in the food or overproduction of acids in the body.

The pituitary and diuresis

The final concentration of the urine by water reabsorption into the circulation occurs in the main collecting ducts. This is intimately controlled by an antidiuretic hormone – ADH – secreted by the posterior lobe of the pituitary gland (pp. 198, 342). This hormone increases the permeability of the duct lining to water. Its secretion is controlled by a feedback mechanism sensitive to the osmotic pressure of plasma and tissue fluids. This is mediated by osmoreceptors in the base of the brain. At normal blood osmolarity, there is a steady receptor discharge and a steady ADH output. Suppose the plasma becomes hypertonic from ingestion of excess sodium chloride; receptor discharge increases, ADH output rises, the duct walls become more permeable, more water is reabsorbed from the urine in the ducts and a smaller volume of more concentrated urine is passed, i.e. there is an *antidiuretic* effect, the object of which is to retain water in the system to dilute excess sodium chloride. On the other hand, excess ingestion of water dilutes the body fluids, ADH secretion is reduced, the duct linings become less permeable and there is a *diuresis*, the passage of a large quantity of pale dilute urine, i.e. the mechanism has functioned to rid the body of its excess water.

The bladder and micturition

Urine is propelled along the ureters by waves of contraction in their muscular walls. It is discharged from the ureteric orifices into the bladder in intermittent jets.

The bladder resembles other hollow organs, as a muscular sac whose outlet is normally closed by a tight ring of muscle, or sphincter. Bladder wall and sphincter must be reciprocal in action; in emptying, the bladder contracts and the sphincter relaxes to allow the efflux of urine, and in resting phases the bladder is relaxed to allow gradual distension as urine collects, while the sphincter remains closed. *Micturition* is the expulsion of urine from the bladder along the urethra. The control of the bladder in this respect is both voluntary and involuntary. Basically, the organ is self-regulating, and empties itself once internal pressure has reached a certain level. When we discuss the vegetative or autonomic nervous system (the independent regulator of automatic functions) we shall see that it consists of two parts, a *sympathetic* and a *parasympathetic* system, with opposite actions but working in coordination. This is so with the bladder also. The parasympathetic nerves empty it by contracting the organ and relaxing its sphincter; the sympathetic allow it to fill by the reverse action.

But there is also a voluntary control which can modify the basic reflexes. There is a second sphincter surrounding the urethra itself, and this is under the control of the will, so that micturition is usually deliberate and aided by voluntary contraction of the abdominal muscles. Nevertheless, once initiated voluntarily, the process continues under automatic control.

The skin and temperature regulation

The skin
The skin is always protective and sensory, but in mammals it is also essential in the regulation of body temperature, for heat loss occurs almost entirely via the skin surface.

Water is lost from the skin by perspiration and sweating. Perspiration is the insensible loss of water vapour; it is continuous, uses the latent heat of vaporisation and produces a loss of weight of almost 0·5 kg in twenty-four hours. Vaporisation is simple physical transpiration, not a function of the sweat glands.

The sweat is an acrid secretion, a weak solution of sodium chloride. The amount of salt lost in a day from profuse sweating may be as much as 113 g, and this is important in the condition known as *heat-exhaustion*, where there is collapse with normal or subnormal body temperature. The condition may be prevented by adding salt to the water drunk. This is quite different from *heat-stroke* (see p. 303).

The sweat glands are controlled by the nervous system, which regulates not the amount of water lost (this is done by the kidneys) but the body temperature. Sweating depends on the difference between internal and external temperatures, the rate of heat production in metabolism, and the humidity of the air, which may be a serious limiting factor if evaporation of liquid sweat is not possible. A minimal rise of body temperature by a third or half a degree sets the mechanism in action. Strangely, the palms and soles are not so stimulated, but they are intimately affected by the emotional state – so-called 'psychic sweating'.

Temperature regulation
The body is maintained at a constant temperature, though the outside level may vary enormously – a particular instance of the maintenance of a constant internal milieu. The body can be kept warm in cold surroundings because heat is produced in oxidative metabolism and cool in hot surroundings by heat loss in sweating. A constant body temperature implies an exact balance between heat production and loss; the latter is a function of surface area and is relatively greater in children, whose surface is much larger than in adults in proportion to body weight.

The average body temperature taken in the rectum is 37·5°C, in the mouth 36·5°C and in the armpit 35·5°C. But there is a daily fluctuation over one or two degrees, highest in the early evening and lowest in the small hours. This fluctuation has superimposed on it, in women, variations at different times of the menstrual cycle. Organs like active muscles are hotter than others such as skin; there is a temperature gradient from the surface to the interior over several degrees.

The regulation of body temperature may be effected by changes in heat production or heat loss; both are simultaneous.

Regulation of heat production
All the tissues produce heat but the muscles predominate; the liver is also an important heat-forming organ as a result of its metabolic activity. With any change in temperature of the environment, heat production is increased or decreased to compensate. In response to cold there is muscular activity, which may amount to shivering; at the same time, liberation of adrenaline stimulates metabolism. External warmth reverses these changes.

These changes are due to the nervous system, which plays on the muscles, and to chemical messengers such as adrenaline. Left to themselves the tissues would behave like those of cold-blooded animals, producing *more* heat as the outside temperature rises; this is what actually happens when the regulating mechanism fails, as in heat-stroke.

Regulation of heat loss

Heat may be lost by *conduction, convection* or *radiation*, or by *evaporation* from the body surface. *Conduction* is of little importance, since the skin and subcutaneous fat are insulators. *Convection* is more important; the air adjacent to the skin is warmed and rises, and each warmed layer is replaced by colder layers, which are warmed in their turn. *Radiation* is also significant since the body surface is a very efficient radiator.

Evaporation varies greatly with circumstances. At low temperatures there is only insensible vapour loss; sweating begins at an air temperature of 29–30°C. At body temperature, sweating and evaporation are the only means of heat loss, since convection and radiation cease when there is no temperature gradient between skin and environment. The cooling effect of evaporation is due to the fact that energy is needed to convert the water of sweat into water vapour, and this energy is abstracted from the body.

About two-thirds of the heat loss of a clothed adult is due to convection and conduction; less than a third by evaporation. These are physical considerations but heat loss is also controllable by physiological activities. Most important is fluctuation in the calibre of the small capillaries of the skin. These may open up to produce flushing, or *vasodilatation*, which increases the surface temperature and so accelerates heat loss. Or they may shut down to produce a cold pallid skin surface – *vasoconstriction* – with reduced heat loss. The maximum blood supply to a patch of skin may be fifty times the minimum. These changes are under the remote control of the central nervous system, via the sympathetic and parasympathetic nerves.

This mechanism ceases to operate when the outside temperature reaches the level at which sweating and evaporation become significant. These are so efficient that a man can stand temperatures that will cook a piece of meat in a few minutes, provided the atmosphere be *dry* so that evaporation is unhindered. If the air is saturated, temperature will rise rapidly, with possible death from heat-stroke. In this condition, there is damage to the vital centres of the brain;

the regulating mechanism breaks down and the body temperature rises until death occurs.

Although sweating and skin blood supply are locally managed by the autonomic nervous system, the ultimate control is from special centres in the brain. These are very sensitive to minute changes in the temperature of the blood and react to restore it to its previous level. They are also sensitive to the *nervous* stimuli that heat or cold produces on the body surface, for the skin is provided with special heat and cold receptors.

The heat-regulating centres work in the ways we have discussed: by producing shivering, sweating, more or less heat production and muscular activity, changes in metabolism, vasoconstriction, or vasodilatation. The mechanism is usually *set* at a certain level and reacts to maintain that level under all circumstances. But if it is upset, the setting may change, and this is what happens in fevers where it is adjusted upwards.

Clothing protects from extremes of temperature by providing a good insulating layer, with air entangled in its interstices. But clothing cuts down the radiating power of the body, and its permeation with moisture is responsible for the chilling effect of a cold, damp climate. Different races solve their heat problems by the way they are made and by their social behaviour.

19

The Nervous System

All protoplasm is *excitable*, and *conducts* excitation, but especially nervous tissue. Nerve cells are particularly sensitive, and their fibres specialise in the transmission of impulses. The nervous system is a network throughout the body, with a two-way connection with central control and permitting a coordinated response to any stimulus from outside or within.

The main parts of the nervous system are: the *central* nervous system (brain and spinal cord) and the *peripheral* nervous system, long bundles of fibres attached to central cells. The *receptor* side of this system conveys information from outside by *afferent* impulses along nerve fibres to the controlling centres, but it also provides information as to the internal state of the body. Some of these messages enter consciousness; others, notably those from within, remain largely unconscious, but do help regulate the working of the body. The *effector* side of the system carries *efferent* messages to the effector organs – muscles and glands – and this response is meant to deal with the situation provoking the original sensory stimuli. From the physiological view, the brain is only the centre of a glorified reflex arc, to which consciousness has been added as an incidental by-product. It does nothing independently, only in response to stimulation from outside.

Basic structure

The basic elements of the nervous system are:

1   The nerve cells or *neurones*;

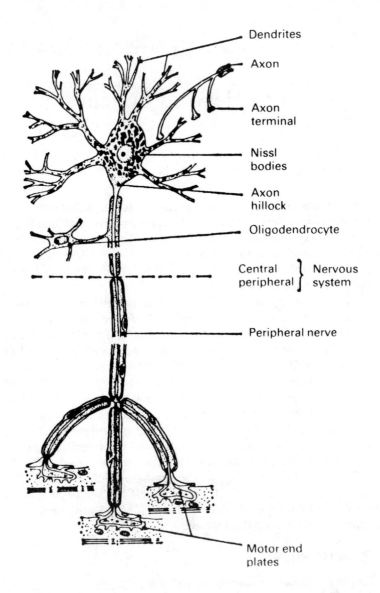

Fig. 19.1　Diagram of a ventral horn cell from the spinal cord (From *A Companion to Medical Studies, Vol. 1*)

Node of Ranvier   Schwann cell nucleus

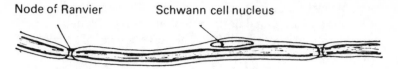

Fig. 19.2 Internodal segment of a nerve (From *A Companion to Medical Studies, Vol. 1*)

2 The supporting cells within the central nervous system, or *neuroglia*;

3 *Connective tissue*, the membranes of brain and cord, and the sheaths of nerve fibres, though there is no connective tissue in actual brain or spinal cord substance.

The neurone (Fig. 19.1)

Each nerve cell has several branching processes, or *dendrites*, interlocking with those of adjacent cells. One is elongated as the main axon for transmission of stimuli; often this is enormously long, traversing the length of the limbs or spinal cord. Cell, dendrites and axon constitute the nerve unit, or *neurone*, and the nervous system is built up of millions of such units in interrelation.

There are two kinds of peripheral nerve fibre: *myelinated* and *unmyelinated*. The large, white or myelinated fibres have a fatty sheath formed by investing Schwann cells, which is indented at intervals at the nodes of Ranvier. These fibres are also found in the

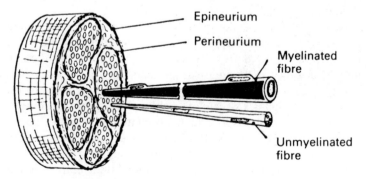

Fig. 19.3 Diagram of the structure of a peripheral nerve containing myelinated and unmyelinated nerve fibres, to show the neural sheaths (From *A Companion to Medical Studies, Vol. 1*)

white matter of the central nervous system. The unmyelinated fibres are finer, and are found in the autonomic nervous system and the *grey matter* of the brain and cord. The nerve bundles of a peripheral nerve contain both kinds of fibre; they are bound together by connective tissue sheaths – the *perineurium* – and the sheath of the entire nerve is the *epineurium*.

Damage to nerve cells is irreparable as they are incapable of reproduction; but the function of destroyed areas in the cerebral cortex may be taken over by other parts of the brain. Injured fibres may grow again, but only in the peripheral nerves.

The nature of nervous transmission

The nerve fibres only *conduct* stimuli, those passing from the central nervous system to muscles or glands, or those travelling from skin and sense organs to the brain and cord. Fibres normally conduct in only one direction, but if we graft a portion of sensory nerve into a motor nerve, the graft will conduct impulses quite efficiently in the reverse direction. Nerves are like telephone wires; the conversation conveyed depends on who happens to be connected at either end. So we associate the optic nerve with sight and the olfactory nerve with smell, but they serve these particular functions only because they link the appropriate sense organ with the corresponding area of the brain; the impulses transmitted are identical to those carried by all nerves.

Nerve fibres are readily stimulated experimentally and stimulation sets up a wave of depolarisation or negative *action potential*, travelling at many metres per second. This wave is not the nerve impulse itself; it is the index of an underlying disturbance of the nature of a sudden change in local membrane permeability to ions.

Nervous action is a stream of small, separate impulses, repeated at very short intervals. There is no continuous flow of nervous excitation, as each impulse is followed by a short *refractory period* during which the nerve will not transmit a stimulus. Further, a single fibre exhibits the *all-or-none* phenomenon; if sufficiently stimulated it will convey an impulse of fixed magnitude, but not a larger or smaller impulse. Nervous tissue undergoes metabolic changes like body tissues generally, using oxygen and generating carbon dioxide with the evolution of heat, though the quantities are minute compared with those concerned with muscular activity. Nervous tissue, particularly the cells of the brain, cannot tolerate oxygen deprivation for more than a minute or two. The large, sheathed motor and sensory fibres of the central nervous system convey impulses at some 90 m/s; in the finer fibres of the autonomic system the speed is only some 1·5–15 m/s. These differences are related to the necessity for immediate response to an external situation; regulation of the internal organs is a more leisurely process.

In a single nerve cell and its fibres, the impulse is conducted uninterruptedly; but there is an inevitable break when the impulse is relayed by another cell. Such relays, or *synapses*, are an integral part of the nervous system. Motor impulses from the brain are relayed by fresh cells in the spinal cord, whose fibres run to the muscles, and sensory impulses from the skin are relayed by cord cells before being transmitted to the brain. The incoming fibre ends in a network of branches which embrace those of the adjacent cell but there is physical discontinuity. Because of this breach, messages are delayed a little at the synapses.

The bridging of the synapse gap is effected by chemical means; the nerve endings liberate chemical substances which stimulate the adjacent cell to start a fresh impulse along its own fibre. A similar gap must be crossed when a motor nerve fibre ends in a muscle, and this again is achieved by liberation of a chemical agent. Although nerve fibres can conduct in either direction, transmission at the synapses is strictly one-way; hence the directional flow in the central nervous system; the same applies to the nerve-muscle junction. Acetylcholine is liberated at motor end-plates in muscle.

The *chemical mediators* are largely acetylcholine or noradrenaline. They are very potent and closely associated with the autonomic system, whose ultimate fibres to glands and viscera are labelled cholinergic or adrenergic as the case may be (p. 328). When we consider the autonomic system, we shall see that chemical agencies play a very large part in regulating internal organs and blood vessels, and that certain endocrine glands (e.g. the adrenals) can secrete the same substances into the blood as hormones to secure a rapid response in emergency.

Anatomy of the central nervous system

The brain and spinal cord, enclosed in the cranium and spinal canal, are continuous at the foramen magnum in the skull base. The twelve pairs of cranial nerves arising from the brain and the thirty-one pairs of spinal nerves arising from the cord constitute the *peripheral nervous system*. Together, they form the cerebrospinal or voluntary nervous system, mainly concerned with control and sensation of the somatic structures of the body wall – skin, muscle, bones and joints – though many of its activities are unconscious.

In contradistinction, the semi-independent *autonomic* or *vegetative nervous system* deals with the automatic functioning of the splanchnic structures – viscera, glands and vessels. Nevertheless, the two systems are closely interconnected.

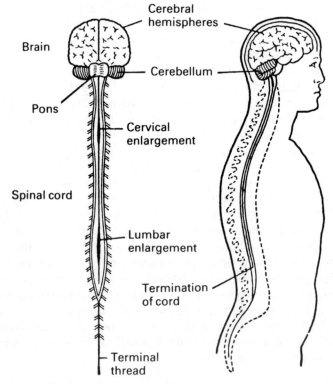

Fig. 19.4   The central nervous system

Cerebrospinal system

Membranes, cerebrospinal fluid

Brain and cord have three enveloping membranes, which are prolonged as sheaths along the nerve roots. The outermost layer is the *dura mater*, a tough loosely applied protective envelope; in the cranium it also forms the lining periosteum of the skull bones. The innermost layer is the *pia mater*, a fine membrane closely applied to the brain and cord, following every cleft and crevice, and carrying with it the fine blood vessels. Intermediate is the *arachnoid layer*; this fits closely inside the dura, but there is a *subarachnoid space* separating it from the pia, filled with cerebrospinal fluid and traversed by spidery connective tissue (Fig. 19.5).

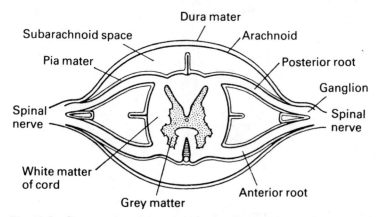

Fig. 19.5 Cross-section of the spinal cord and its membranes (After *Gray*)

The cerebral and spinal membranes are continuous, the fluid bathing the outer surfaces of brain and cord. It also occupies the hollow chambers or ventricles of the brain which communicate with the central canal of the cord. It is secreted by the vascular *choroid plexus* lining the ventricles, circulates inside the brain and cord, escapes through the roof of the hindbrain to the subarachnoid space and is reabsorbed into the bloodstream via the venous sinuses of the cranium. There is a continuous circulation of fluid from the blood vessels in the ventricles, round the brain and cord, and back into the bloodstream.

Spinal cord

The spinal cord is elongated and cylindrical with two swellings, the *cervical* and *lumbar enlargements*, which are the origins of the roots of the brachial and lumbar nerve plexuses for upper and lower limbs.

In the foetus the cord occupies the length of the spinal canal, but it fails to keep pace with growth, and in the adult, ends at the first lumbar vertebra in a conical extremity. Since the nerve roots must still emerge at the correct levels from the intervertebral foramina, they have to run more and more obliquely downward and out as they arise lower in the cord; so the spinal canal below the termina-

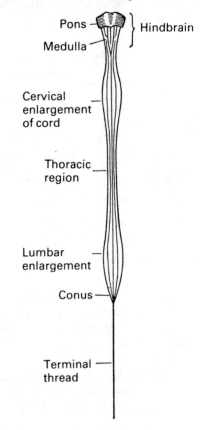

Fig. 19.6   Spinal cord and hindbrain (After *Gray*)

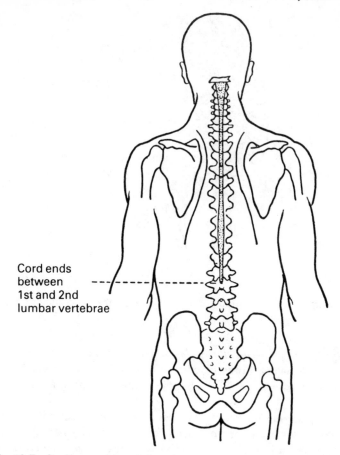

Cord ends
between
1st and 2nd
lumbar vertebrae

Fig. 19.7 Surface marking of the spinal cord (After *Gray*)

tion of the cord is filled with a mass of roots – the cauda equina or 'horse's tail' – descending to their lumbar and sacral exits.

Cross-section of the cord shows the inner H-shaped arrangement of *grey matter*, composed of nerve cells, and the surrounding *white matter* – descending and ascending tracts of nerve fibres. In the very centre is a minute canal. In the midline a cleft anteriorly and a fissure posteriorly divide the cord into symmetrical right and left halves.

On each side the grey matter projects in front and behind as the

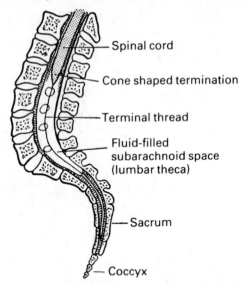

Fig. 19.8   The termination of the spinal cord in the upper lumbar region (After *Gray*)

anterior and *posterior horns*. The *anterior horn* contains the motor nerve cells, whose fibres are destined for the stimulation of muscles, and these leave the cord in a bundle as the anterior or motor nerve root.

The *posterior horn* contains sensory cells; fibres enter it from the posterior or sensory nerve root and are mediated by a group of cells forming a knob or ganglion on this root just outside the cord (see Fig. 19.9).

Spinal nerves
The two roots on each side join just beyond the ganglion to form the spinal nerve proper, the junction lying within the intervertebral foramen before the nerve leaves the spine. Thus the spinal nerve contains sensory and motor fibres, and these are not separated again until, in the nerve branches in limb or trunk, specific motor and sensory twigs are given off to muscles and areas of skin.

The simplest pattern is in the thoracic region, the spinal nerve on each side encircling the chest in its appropriate intercostal space

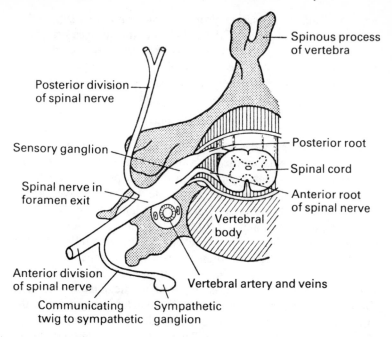

Fig. 19.9 Cross-section showing the spinal cord, its issuing roots and the appropriate spinal nerve, as they lie in relation to the bony spine. The section is in the cervical region. Note the communicating twig to the sympathetic ganglion (After *Gray*)

without connection with its neighbours. But the lower cervical and the lumbar and sacral nerves form complex networks as soon as they have left the spinal column. From these plexuses emerge the peripheral nerves of the limbs, e.g. median and ulnar in the arm, sciatic and femoral in the leg; these are usually mixed motor and sensory, though some are almost entirely one or the other.

Our cross-section of the cord is much the same whatever the level of the section, i.e. the cord consists of a number of identical *segments*, to each of which is attached a pair of spinal nerves. This is a relic of primitive segmentation, a pattern obscured in man by the development of head and limbs, which lingers on in the repetitive arrangement of ribs, vertebrae and the structure of the cord itself. It is important because each cord segment controls the sensation of a particular area of skin and the movement of a particular group of muscles. Thus the sensation of the little finger is always supplied by the first thoracic segment of the spinal cord, the muscles straightening the knee always derive their motor nerve fibres from the third and fourth lumbar segments, and so on.

This is of great value in helping to locate the precise site of injury or disease of the central nervous system.

The reflex arc

The spinal cord provides a relatively simple means for performing essential actions without the intervention of the brain. An irritation of the skin sends a sensory impulse through the posterior root to the posterior horn of the appropriate segment; this is relayed to the anterior horn cells of the same or related segments and becomes a motor impulse travelling out through the anterior root to activate muscles which withdraw the affected part. This is the reflex arc at its simplest, the so-called withdrawal reflex. It is independent of the brain, though the brain is aware of its results and can modify them. Another simple reflex arc is involved in the *tendon-jerk*. The simplest example is the knee-jerk; when one leg is crossed over the other, a tap on the patellar tendon (Fig. 7.9) causes the quadriceps muscle to straighten the knee. There are many similar reflexes, based on the stimulation of stretch receptors in the muscle.

More complex, but basically similar, are the reflexes of micturition, defaecation, orgasm and childbirth, all controlled by a particular segment or group of segments. All these activities remain possible, even when the cord-brain link has been severed, so that the body is paralysed as regards conscious sensation and voluntary movement.

The tracts of the cord

The grey matter is composed mainly of cells. The white matter is made of fibres running longitudinally in cable fashion, grouped in a constant pattern of individual tracts, some sensory and some motor. A few tracts run the length of several segments only, but most run from cord to brain and link the anterior or posterior horn cells with the appropriate centres of the brain. Thus, from the motor area of the cerebral cortex concerned with arm movement, fibres run to the anterior horn cells of the seventh cervical segment of the cord, and from these fibres leave to join the brachial plexus, enter the median nerve and actuate the flexors of the wrist. Sensory impulses from the knee travel up the femoral nerve, through the lumbar plexus to the posterior horn cells of the fourth lumbar segment, and are relayed up the long sensory tracts to the sensory portion of the cerebral cortex.

In other cases, other parts of the brain are the termini – cerebellum or midbrain – but an invariable principle is that of crossing or *decussation*; at some point, usually in the brain stem, the tracts cross from one side to the other so that the sensory and motor regions of the brain control the opposite halves of the body.

The brain
The brain is the elaborated upper end of the cerebrospinal axis. It has the same membranes as the cord, it almost entirely fills the cranial cavity and it makes indentations on the inner aspect of the cranial bones. The grey matter is now on the surface and the white fibre tracts within. The main portions of the brain are, from above down, as follows:

1 The *forebrain* is the great overhanging pair of *cerebral hemispheres* – the bulk of the organ – symmetrical, rounded masses of convoluted nervous tissue which hide the lower parts of the brain when viewed from above.

The two hemispheres are separated by a deep fissure, with a partition of dura mater, but are connected at the base of this cleft by a great bridge of fibres. The superficial grey matter, the cerebral cortex, has an enormous area owing to its intricate convolutions. This produces a number of ridges, or *gyri*, separated by valleys, or *sulci*.

Each hemisphere is composed of several *lobes* which are not clearly demarcated. In side view, the *frontal lobe* is seen at the anterior end of the hemisphere, in the anterior cranial fossa. Behind is the *occipital lobe*, in the posterior cranial fossa. The *temporal lobe* lies behind the frontal and projects below, occupying the middle cranial fossa, and the *parietal lobe* is an ill-defined area between frontal and occipital regions above the temporal lobe.

The main motor and sensory areas are midway between the poles, and the body is represented in them in inverted fashion, i.e. the areas for the legs are uppermost, those for the head below. There is

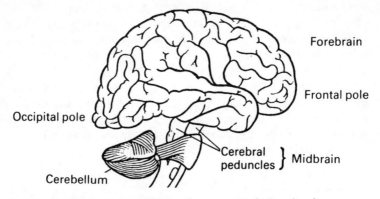

Fig. 19.10   The brain in side view; diagrammatic (After *Gray*)

a vaguely localised centre for the appreciation of the meaning of words in the temporal lobe of the left side; this is on the right side in left-handed people. All these functional areas are concerned with the opposite half of the body, though the visual area of the occipital cortex is concerned not so much with the opposite eye as with gathering all the impulses from the opposite field of vision that have entered both eyes. A considerable area of cortex is assigned no particular function. These are the psychic or association areas, concerned with correlation of data, and with the higher levels of personality, particularly in the frontal lobes.

Cross-section of the forebrain shows the pattern of sulci and gyri, the contrast of grey and white matter, and the cleft between the hemispheres (Fig. 9.11). It shows on each side the hollow *ventricle* in the depths of each hemisphere, filled with cerebrospinal fluid, and great masses of grey matter beside the ventricles, the *basal ganglia*, one of which – the *thalamus* – is connected with the emotions.

2   The *midbrain* is situated at the base of the hemispheres and consists partly of a squat *peduncle* on each side, a pillar supporting the corresponding hemisphere and carrying fibres to and from it. The roof of the midbrain contains the nuclei of the third cranial nerve, which controls movement of the eyeballs.

The *pituitary body* projects from the undersurface of the brain in front of the peduncles.

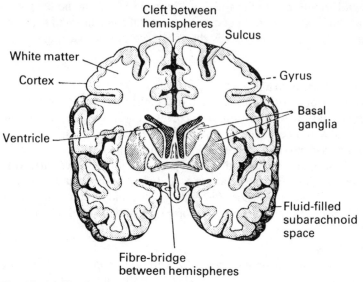

Fig. 19.11 The forebrain in cross-section

3 The *hindbrain* is best seen on the underside of the organ (Figs 19.6, 19.10, 19.11). It includes:

(a) The *medulla oblongata*, the bulbous upward prolongation of the spinal cord, the lowest part of the brain, containing the nerve centres for heartbeat and respiration;

(b) The *cerebellum*, a pair of rounded hemispheres, finely convoluted, occupying the deepest part of the posterior cranial fossa;

(c) The *pons*, a broad bridge of fibres connecting the cerebellar hemispheres.

The hindbrain is the most primitive part of the organ, and is concerned with the basic activities of the body. It is independent of conscious control and continues to function in sleep, unconsciousness or deep anaesthesia. The medulla's vital centres regulate heartbeat, respiration and blood-pressure, and remain active until the last moment before death. The cerebellum is concerned with the regulation of muscle tone and posture, in response to the sensory impulses of position and tension from joints and tendons, and also to information as to the body's position in space from the semicircular canals of the ear. It governs the postural mechanism of head and trunk, the righting reflexes that preserve upward carriage of the head.

Fibre tracts of brain

A series of cell-stations is situated at higher and lower levels in the grey matter of the cerebral cortex, basal ganglia, midbrain, medulla and cerebellum. To and from these great tracts of nerve fibres ascend and descend within the white matter, the axial core of medulla and midbrain. These tracts were encountered in cross-section of the spinal cord, and may be classified as motor (descending) and sensory (ascending); at some level, usually in the hindbrain, they all cross to the opposite side. The principal tracts are:

Motor tracts

1 *Pyramidal*, running from the cerebral motor cortex to the anterior horn cells of the cord at all levels, thence relayed to the motor roots; concerned with voluntary motion;
2 *Cerebellospinal*, running from the cerebellum to the anterior horn cells, concerned with automatic regulation of tone and posture.

Sensory tracts

1 The posterior bundles of the cord carrying touch and sensation to the cerebral sensory cortex;
2 The anterior and lateral bundles of the cord, carrying pain and temperature sensation to the sensory cortex;
3 The spinocerebellar bundle carrying postural sensation to the cerebellum.

The sensory bundles relay impulses entering the cord via the posterior sensory roots and posterior horn cells. Since they acquire an increasing influx as they ascend, and since the motor tracts shrink as they descend, because they shed fibres to successive segments, the cord as a whole tapers from above downwards.

The ventricles

The brain is hollow and traversed by channels for the cerebrospinal fluid. In each cerebral hemisphere is a large *lateral ventricle*, and between and below these a small *third ventricle*, from the floor of which the pituitary body is slung. These communicate with a tiny channel, the *aqueduct*, which leads back through the midbrain to the *fourth ventricle* of the medulla. This is continuous with the central canal of the spinal cord, and from its roof the fluid escapes to bathe the outer surface of the brain and cord in the subarachnoid space. Growing from the linings of the lateral and fourth ventricles is the vascular *choroid plexus*, which secretes the fluid; any block to its free flow results in a damming-up, which balloons the brain to produce the condition known as *hydrocephalus* or 'water on the brain'.

The cerebrospinal fluid cushions the brain and cord against concussion or violent changes of position. It is normally maintained at a constant pressure which is related to the pressure of blood in the great veins and is raised when venous tension is increased by coughing or straining. Any disease process or tumour within the unyielding cranium increases the cerebrospinal fluid pressure as it expands, causing headache, and vomiting.

The composition of the fluid and its contained cells may be affected by disease, so examination of the fluid is often of value in diagnosis. While the cord ends at the first lumbar vertebra, its membranes continue as far as the sacrum. The lumbar spine thus encloses a space, or *theca*, filled with cerebrospinal fluid, and this may be entered by a long needle inserted between the spinous processes – the procedure of *lumbar puncture*.

Vessels and nerves of the brain

The main *arterial* supply of the brain is derived from the two internal carotids, which enter the skull base. They form, with the vertebral arteries which have entered the foramen magnum, a vascular circle around the stalk of the pituitary body. From this, *anterior, middle* and *posterior cerebral arteries* are given off to the frontal, parietotemporal and occipital parts of the brain. The hindbrain is supplied by the *vertebral artery*.

The main *veins* form sinuses, which run between the layers of dura mater and drain to the internal jugular; the *superior sagittal sinus* of the vault, which runs the length of the brain from front to back between the hemispheres, also drains off the cerebrospinal fluid, which is constantly reabsorbed.

The brain substance has no nerves; it can be handled without reaction, even in the conscious patient. But its covering membranes are extremely sensitive.

The cranial nerves

In certain parts of the brain there are collections of nerve cells, forming the *nuclei* from which spring the twelve pairs of cranial nerves, best seen on the underside. These are concerned with movement and sensation in the head and face, including the eye, ear and nose; but one (the eleventh) controls certain neck muscles and another (the tenth or vagus) is distributed to the internal organs of the neck, chest and abdomen. They may be listed from above downwards:

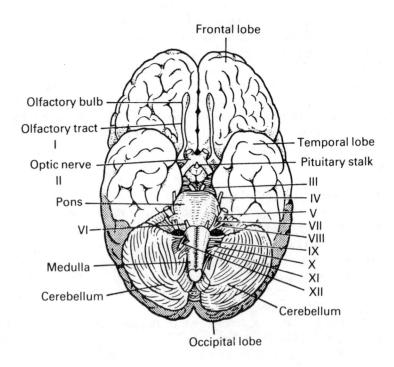

Fig. 19.12 The underside of the brain, showing the attachments of the twelve pairs of cranial nerves

Cerebrum

(I) The first or *olfactory* nerves are those of smell. On each side a group of some twenty rootlets ascends from the nose, through the perforations in the floor of the anterior cranial fossa, and ends in the olfactory bulb under the frontal lobe. From this, the olfactory stalk relays the impulses to the cerebral hemisphere. These nerves are purely sensory.

(II) The second or *optic* nerve is concerned with sight. Each carries fibres from the nerve cells of the retina and leaves the back of the orbit as the stalk of the eyeball through the optic foramen to enter the middle cranial fossa. The two nerves form an X-shaped crossing or *chiasma* under the frontal lobes, in front of the pituitary stalk, the hind limbs of the X being the *optic tracts*. The effect of this is that fibres from the outer half of each retina remain in the optic tract of the same side, while those from the inner halves cross into the opposite tract. Thus the right tract carries all the fibres from the right halves of both retinae and, therefore, all the sensation from the left half of the visual field, and *vice versa* for the left tract. This separation is essential to maintain one-sided representation of function in each half of the brain. The eye is a one-sided organ, but sees on both sides of the body, and what is needed for coordination is that each side of the *visual field* should be represented as such in the cerebral cortex. The impulses are relayed to the visual area of the occipital cortex, and are connected with the nucleus of the third nerve, which controls eye movements. The optic nerve is purely sensory.

Midbrain

(III) The third or *oculomotor* nerve controls four of the six small muscles moving the eyeball, and is purely motor.

(IV) The fourth or *trochlear* nerve controls another of the ocular muscles, and is purely motor.

Hindbrain

(V) The fifth or *trigeminal* nerve has a large sensory root divided into three branches:

> *Ophthalmic*; this conveys sensation from within the orbit other than sight, and sensation from the forehead and front of the scalp;

Maxillary; this conveys sensation from the upper jaw and teeth and overlying skin;

Mandibular; this conveys sensation from the lower jaw and teeth and overlying skin, and ordinary sensation from the front of the tongue and mouth.

The mandibular division enters a canal in the lower jaw. The ophthalmic division enters the back of the orbit. The maxillary division runs in a canal between the orbital floor and the roof of the maxilla to emerge on the front of the cheek.

The fifth nerve also has a small *motor root*, concerned with the muscles of mastication. Thus it is a mixed sensory and motor nerve.

(VI) The sixth or *pathetic* nerve controls the last of the ocular muscles. It is purely motor.

(VII) The seventh or *facial* nerve supplies the facial muscles; it also activates the secretion of some of the salivary glands and conveys taste sensation from the front of the tongue. It leaves the posterior fossa of the skull by entering the internal auditory meatus on the inner aspect of the temporal bone with the eighth nerve. After a winding course, it emerges in front of the mastoid process, and enters the parotid gland, in which it divides into twigs supplying the facial muscles and the platysma in the neck. It is a mixed sensory and motor nerve.

(VIII) The eighth or *auditory* nerve connects the inner ear, deep in the temporal bone, with the hindbrain; it serves two distinct functions:

(*a*)   The *acoustic portion* carries sensations of sound and pitch from the organ of hearing, the cochlea; these are relayed to the auditory part of the cerebral cortex;

(*b*)   The *vestibular portion* carries postural sensation from the semicircular canals of the internal ear to the cerebellum. It provides information about position.

(IX) The ninth or *glossopharyngeal* nerve is mainly sensory; it also activates parotid secretion and some little muscles of the pharynx. It conveys sensation from the pharynx, tonsil and back of tongue, including taste.

(X) The tenth nerve, the *vagus*, is very important, extending through the neck and thorax to the abdomen. It is an essential component of the autonomic system controlling the viscera. Its anatomical course in neck and chest was described on p. 211. Its main branches are:

(*a*) In the *neck*, for movements of the pharynx and larynx, sensation of their linings.

(*b*) In the *chest* it forms the cardiac and pulmonary plexuses; it also supplies the muscle of the oesophagus and bronchi, and their associated glands, and carries sensation from their linings;

(*c*) In the *abdomen* it supplies the muscle, glands and mucous membranes of the stomach and duodenum, and sends twigs to the liver, spleen and kidneys.

(XI) The eleventh or *accessory* nerve is purely motor and supplies the muscle of soft palate and pharynx, and the sternomastoid and trapezius.

The ninth, tenth and eleventh nerves all arise from the medulla and leave the skull base through the jugular foramen with the internal jugular vein.

(XII) The twelfth or *hypoglossal* nerve controls the muscle of the tongue and floor of the mouth.

The fibres of motor cranial nerves and those of the motor roots of the cord arise from their cells in the cerebral nuclei or anterior horns and pass without a break to their destined muscles. But sensory fibres, passing in from the skin or other sensory surface, end at a group of cells just outside the brain or cord; these make up the *ganglia*, knobby swellings on the posterior spinal roots, or on the sensory cranial nerves at the skull base. All sensory nerves have a ganglion, and it is the axons of the ganglion cells, not the original sensory fibres, that enter the cerebrospinal axis. In other words, motor impulses are direct, sensory impulses are mediated. The outgoing motor pathways are known as efferent, the incoming sensory fibres are afferent.

Autonomic nervous system

Sympathetic and parasympathetic (Fig. 19.13)
The autonomic nervous system is distributed to the muscular coat of the blood vessels, and to the smooth muscle and glands of the

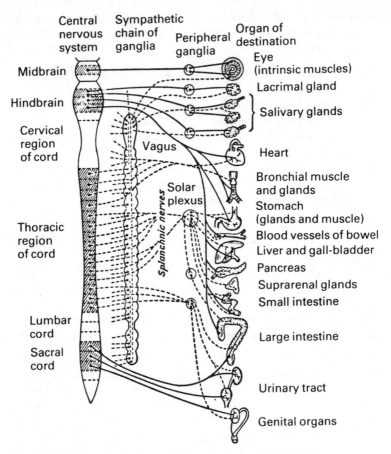

Fig. 19.13   Plan of the autonomic nervous system (After *Gray*)

viscera. It controls their movements and secretion, and monitors their distension. All its nerve fibres are ultimately derived from the cerebrospinal axis itself, and it is susceptible to modification from the centre, not by an act of will but through the emotions.

Anatomical arrangements
The *sympathetic system* is a chain of ganglia on each side of the spinal column from atlas to coccyx, where the two meet and fuse in a single ganglion. We have met this chain lying on the sides of the

vertebral bodies at the back of the abdominal and thoracic cavities, where there is one ganglion to each segment and spinal nerve, and in the neck where there are only three. The outflow from the central nervous system to the sympathetic is by a little communicating twig from each spinal nerve in the thoracic and upper lumbar region to the corresponding sympathetic ganglion (Fig. 19.9). From this ganglion the peripheral sympathetic system develops in two ways:

1   By a returning twig to each spinal nerve, the sympathetic fibres travelling out with that nerve to the somatic structures of the limbs and body wall;
2   By the formation of *plexuses*, e.g. the cardiac and pulmonary plexuses of the chest, branches of which form a network around the vessels to heart and lungs; also the splanchnic nerves run down from the lower thoracic chain through the diaphragm to form the coeliac plexus around the upper part of the abdominal aorta, from which branches go to the viscera.

The *parasympathetic system* is a less distinct structure, for its fibres are contained within certain cranial and sacral nerves at either end of the cerebrospinal axis – the craniosacral outflow. The former supply the glands and vessels of the head and eye, the latter the bladder, rectum and genital organs. But the tenth cranial nerve – the *vagus* – is the great parasympathetic nerve of the body, traversing neck, chest and abdomen to supply the viscera as a whole.

There is a major difference between the autonomic and central nervous systems. In the latter, the cell bodies of the neurones lie *within* the system; only fibres lie without the brain and spinal cord. In the autonomic, there are peripheral ganglia, consisting of masses of cell-stations with synapses. In the case of the sympathetic, these are located close to the vertebral bodies; in the parasympathetic, they are near or in the organs. So we speak of preganglionic and postganglionic fibres in the autonomic system. The parasympathetic has long preganglionic fibres in a spinal nerve before synapsing with ganglion cells close to the organ they supply, so the postganglionic fibres are very short. Conversely, the sympathetic outflow has short preganglionic fibres, but their postganglionic course is often very long.

Chemical transmission in the autonomic system takes place (*a*) at the ganglionic synapse, and (*b*) where the postganglionic fibres enter the target organ. In the former site, both sympathetic and parasympathetic act by liberating acetylcholine. At the peripheral site, the sympathetic liberates acetylcholine where sweat glands and skeletal muscle blood-vessels are concerned, but noradrenaline in the heart, visceral muscle, glands and internal vessels; the parasympathetic again liberates acetylcholine. The actions of noradrenaline and acetylcholine are completely antagonistic; noradrenaline is the chief sympathetic transmitter at the end-organs.

The effects of sympathetic stimulation are those of bodily activity in relation to fear, fight or flight. Aided by adrenaline secreted by the adrenal gland, it constricts the blood vessels of the skin, so raising the blood-pressure and shunting blood to heart and brain; speeds and strengthens the heartbeat; dries up glandular secretion; dilates the pupils; stands the hair on end and initiates sweating; and relaxes the walls of the hollow viscera. In contrast, the parasympathetic produces relaxed and constructive activities in tranquillity, as after a heavy meal. It dilates the peripheral vessels, slows the heart and lowers the blood pressure, and excites secretion and peristalsis.

However, this is an oversimplification. The two parts of the system always work together to control the internal organs, even though at any time one or other may predominate.

Special sense organs

Sensation in general

Information from the muscles, joints and internal organs is constantly arriving at the brain; but, except pain, these messages do not enter consciousness. The special sense organs – the eye, ear, nose, tongue and skin – provide information of a particular quality, and elicit complex responses. The particular nature of a sensation is determined by the receiving organ, not by the stimulus. The same cause will produce different effects when applied to different sense organs – e.g. radiation may be appreciated as light by the eye but as warmth by the skin, a tuning fork is only a touch to the skin, but conveys a note to the ear. Conversely, a sense organ can only react in one way, whatever the stimulus; the eye not only reacts to light normally but also transmits the sensation of light when it is pressed vigorously.

Another characteristic of the relation between brain, sense organs and the outer world is that of *projection*. Although sensations are only appreciated as the result of changes in the brain cells, they are felt as if taking place more remotely; touch is projected to the skin and taste to the mouth, while we regard sight and sound as coming from the surroundings.

Finally, sense organs obey physical laws. There is a minimum threshold value for the stimulus, below which it is ineffective, and this level depends on physiological factors; it is higher in fatigue or if there are distracting stimuli. Further, we do not experience sensations in an intensity directly proportional to the physical intensity of the stimulus; there is such an enormous range of impressions from the outside world that we have to summarise them. A light that we feel to be twice as bright as another is, by physical measurement, perhaps ten times as intense, and this relation applies also to sound, i.e. it is logarithmic.

The important factor in appreciating a stimulus is the change from a steady state. The fresh stimulus causes a burst of impulses, which die away, even though the stimulus persists – the phenomenon of *adaptation*. This is essential, or we could not tolerate the continued stimulus of sunlight, or the pressure of our clothes.

The skin as sense organ

The skin intervenes between the body and the outside world, so its sensory reactions are of major importance. In fact, the eye and ear, and the central nervous system itself, have developed from an infolding of this surface layer. The modalities involved include touch, pressure, pain, temperature, vibration and stereognosis.

Touch is transmitted by small end-bulbs of the cutaneous nerves, in little bays on the deep surface of the epidermis (Fig. 19.14). The hairs play an important part by magnifying the effects of contact, and the sensory bulbs are grouped around their roots; hence the phenomenon of *tickle*, and the blunting of sensation that follows shaving a hairy surface. The skin is not uniformly sensitive; some points are more sensitive than others. There are a number of scattered *touch spots*, separated by ½ mm or so, and between these there is no sensation. The capacity of discriminating between

Skin surface

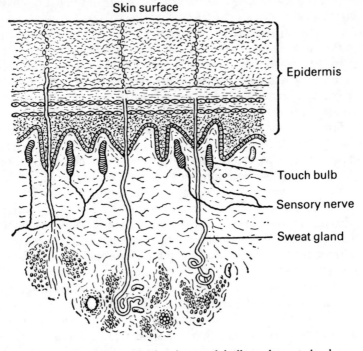

Epidermis

Touch bulb

Sensory nerve

Sweat gland

Fig. 19.14   Section of the skin showing touch bulbs and sweat glands

adjacent stimuli varies from place to place. On the tongue and fingers, where touch spots are close together, two compass points are felt to be separate if only 1 mm apart, but they must be separated by 6 cm on the thigh before being appreciated as distinct.

The skin illustrates how our conventional reading of sensation depends on habituation. A pencil point between the tips of middle and index fingers is felt to be single because these surfaces are normally adjacent and we have learnt to fuse their sensations; if the fingers are crossed, surfaces that are never normally in contact are brought together and a pencil placed between them is now felt as double.

It is now doubtful whether there are specific *pain* receptors and fibres. Pain and touch are intermingled; for instance, in the cornea of the eye, any touch is felt as pain, and there is no ordinary touch sensation. The object of pain sensation is to protect by provoking an automatic withdrawal reflex. When pain sense is lost in nervous disease, the sufferer may burn himself and remain unaware, and the skin becomes liable to infection and ulceration.

There is a close link between temperature and pain, and both are transmitted along the same fibre tracts in the central nervous system. The headquarters for pain appears to be the thalamus, and 'pain' seems to be the result of a varying analysis of stimuli, for the same sensation is sometimes painful and sometimes not, and pain is always affected by attention and emotional state. The severance of the fibre tracts connecting frontal cortex and thalamus may render even severe pain tolerable.

The viscera are normally insensitive to cutting or burning. The brain itself appears to be entirely insensitive. Then there is the phenomenon of *referred pain*, where pain originating from a deeper structure is felt over the skin surface. Thus pain from the diaphragm is often felt at the shoulder because the phrenic nerve originates from the same segments of the spinal cord that supply the skin of the shoulder; and a prolapsed intervertebral disc in the lumbar region irritates a root of the sciatic nerve and causes pain in the leg.

Temperature sensation
Feelings of heat and cold are due to a loss or gain of heat from the object felt. Temperature sensation is not distributed uniformly but in discrete points at the skin surface, and there seem to be separate sets of receptors – heat and cold spots. The mucous membranes are less sensitive to heat, so that one can drink fluids that are too hot to be held. It is a matter of observation that there is a qualitative difference between the hot and the merely warm, but the problem of temperature sensation is complicated by the changes produced in the small blood vessels of the skin. If these open up, the part becomes warm because flushed with blood, while constriction results in coldness. A painful stimulus often causes vasodilatation and is spoken of as a *burning* pain.

The skin is also sensitive to *vibration* – more so if the vibration is transmitted to underlying bone. Vibration is closely linked to touch and pressure sensation.

Finally, there is *stereognosis*, the ability to recognise three-dimensional shapes. For this, the object has to be handled, not merely felt, the fingers must be warm and sensitive, and the cerebral cortex must be able to analyse the information supplied.

The eye
The globe of the eye lies embedded in the orbital fat, embraced by six small *ocular muscles* that spring from the walls of the orbits and move the eyeball in all directions. It is a sphere almost 2·5 cm in all diameters, with a more local bulge at the anterior pole formed by the window of the *cornea*. The *optic* nerve is attached behind and runs back like a stalk to reach the cranial cavity via the optic foramen.

The eyeball has three coats:

1   An outer fibrous coat, the tough opaque *sclera*, the white of the eye, which surrounds the globe except where it merges into the cornea.
2   An intermediate pigmented layer, the *choroid*, which lines the sclera and is modified in front to form the diaphragm of the *iris*. This rests on the lens and has the central aperture of the *pupil*, the size of which is varied by contraction of the underlying muscular *ciliary body*.

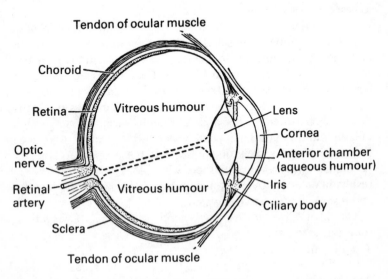

Fig. 19.15   Section through the globe of the eye

3 The innermost nervous layer is the *retina*, adapted for the reception of light stimuli and extending forward no further than the ciliary body. It is pigmented, containing the visual purple which is bleached by exposure to strong light; and consists of a layered pattern of nerve cells, whose fibres leave the eye in the optic nerve. The point of attachment of the latter to the retina is the *optic disc* or blind spot; here the central retinal artery and vein enter the eye by traversing the optic nerve.

There are three refracting media of the eye:

1 The watery *aqueous humour* fills the anterior chamber, the space between the cornea and lens, and also the little posterior chamber, the narrow cleft between lens and back of iris. Like the cerebrospinal fluid, the aqueous is constantly secreted and re-absorbed.

2 The translucent *lens* is a solid structure enclosed in a capsule, more convex on its posterior aspect. This curvature is modified by changes in the tension of its suspensory ligament to accommodate for near and far vision.

3 The bulk of the eye is filled with the thin translucent jelly of the *vitreous humour*, traversed by a central canal running from the entrance of the optic nerve to the back of the lens.

Eyelids and lacrimal apparatus

The supporting substance of each lid is the tarsal plate, a layer of dense connective tissue; the eyelashes are attached at the lid margins, and the inner surfaces are lined by a delicate membrane, the *conjunctiva*, which is reflected over the cornea and sclera. These are kept moist by the secretion of tears by the *lacrimal gland*, situated in the upper and outer part of the orbital cavity and with ducts opening under the upper lids. The secretion is collected in the *lacrimal sac* at the inner angle of the lids; from here the *nasolacrimal duct* runs down the sidewall of the nasal cavity to discharge into the nose.

The eye as an optical system

We may compare the eye to a camera, to the former's advantage. Focusing is automatic; and the lens changes its shape, or *accommodates*, to become more curved and powerful for near vision and less so for distant objects, the aim being to converge the rays of light exactly on the retina. A normal eye produces a sharp retinal image of an object at infinity without any accommodation at all. There is a very wide field of vision, over 200° of a circle, so that objects slightly behind eye level are still in sight, and there is little loss by reflection at the refracting surfaces.

The cup-shaped retina is in just the position for the best definition of the image, and it has two different systems of receptor cells: rods and cones. The rods, with their visual purple, are for blacks, whites and greys under twilight conditions. The cones have other pigments for colour vision in bright light; when this system is defective, the individual is colour-blind. In darkness the retina becomes increasingly sensitive to the available light, the phenomenon of *adaptation*.

Refractive errors
Longsightedness is normal in early childhood, for the eyeball is too small for proper convergence. If this persists, as *presbyopia*, in adults, light rays converge behind the retina and correction is by a convex spectacle lens to supplement the converging power of the natural lens. The long sight of old age is due to loss of resilience of the lens so that it cannot assume a more convex shape.

In shortsightedness, or *myopia*, rays are focused in front of the retina by too powerful refraction. This is corrected by a supplementary concave lens to displace the point of convergence backwards.

Astigmatism results from the lens/cornea system being more curved in one plane than another, i.e. not truly spherical. Accommodation is correct for a given line, but a line at right angles is not properly focused. Correction is with a cylindrical lens.

Visual acuity is not a measure of the *size* of an object, for anything is visible if it emits enough light. It is measured by estimating the minimum angle subtending two lines that can be appreciated; normally, this is one minute of arc.

Binocular vision is important because the combined use of both eyes gives a wide field of vision and eliminates the screening effects of nose and brow. It also provides the stereoscopic effect, which depends on the fact that the images of an object formed by each eye are slightly different, but are presented simultaneously to the brain without appearing double. To secure this, both eyes must converge

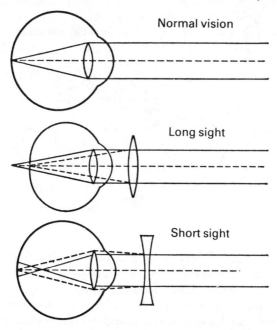

Fig. 19.16 The focusing of parallel light rays by the optic lens in normal vision, and in long and short sight. When there is a refractive error an artificial lens has to be used to help the rays converge on the retina; the dotted lines show how this is achieved

on the object. In individuals with poor converging power, the single image is easily dissociated into two; and, if there is a habitual preference for a better eye, the image of the weaker one is permanently suppressed and this eye may become virtually blind.

The ear

The ear has three parts: the *external ear*, the *middle ear* or *tympanic cavity* and the *internal ear* or *labyrinth*.

The *external ear* consists of the *pinna*, the funnel-shaped organ for collection of sound waves, a thin plate of elastic fibrocartilage covered with skin. This leads along a narrow channel, the *external auditory meatus*, 2·5 cm long, cartilaginous at the surface but bony as it approaches the skull, and lined by skin with wax-secreting glands. The external ear is mostly outside the skull, while the middle

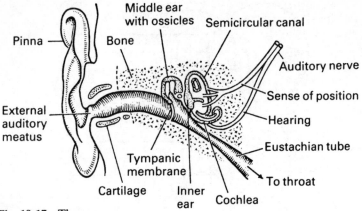

Fig. 19.17 The ear

and inner compartments are within the temporal bone. At the bottom of the external meatus, the circular *tympanic membrane –* the eardrum – shuts off the middle ear cavity.

The middle ear is a roughly cuboidal cavity. Its roof is the floor of the middle cranial fossa, the eardrum is its outer wall and the bony case of the inner ear lies medially. A canal, the *Eustachian tube*, connects it with the nasopharynx. Its function is to keep the air pressures identical on either side of the drum; if the pressure is suddenly altered, as in an aeroplane ascent, there is temporary deafness until the canal is opened by yawning or swallowing. The cavity is spanned by three tiny bones, the *auditory ossicles*, which link the tympanic membrane with the outer wall of the inner ear, transmitting by bone conduction the vibrations set up by sound impulses in the eardrum.

The *internal ear*, deep in the temporal bone, is a complex organ, the *membranous labyrinth*, housed within a bony chamber. The upper part is concerned with sense of position and spatial orientation; this consists of three fluid-filled *semicircular canals* arranged in planes mutually at right angles. Then there is an immediate portion, the *utricle and saccule*. And below is the *cochlea*, the true organ of hearing, a spirally-coiled structure responding to pitch.

On the inner cranial side an *internal auditory meatus* allows the passage of the seventh and eighth cranial nerves between the temporal bone and the brain.

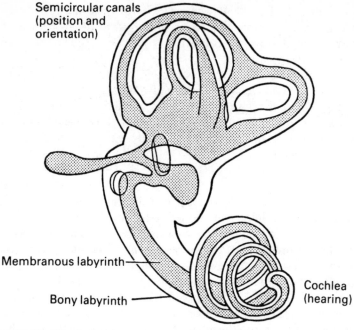

Semicircular canals
(position and
orientation)

Membranous labyrinth

Bony labyrinth

Cochlea
(hearing)

Fig. 19.18 The internal ear

Gravity and position sense

The semicircular canals, utricle and saccule, are filled with fluid and
lined with hair cells, which respond to changes in position of
jelly-like mineralised masses – the *otoliths*, which respond to gravi-
tational changes brought about by changed positions of the head.
Rotational movements of the head stimulate the hair cells of the
canals. Control of accurate movements of the eyes to compensate
for head movements depends on information supplied from the
canals. Motion sickness is due to overstimulation of the canal
system. Impulses from the canals are transmitted along the vestibular portion of the eighth cranial nerve to the cerebellum, which
adjusts posture in response.

Hearing

The tension of the eardrum is not constant but is adjusted by a
muscle which springs into contraction to protect the delicate middle
ear mechanism against very loud noises. The membrane responds

faithfully to high and low frequencies between 40 and 30000 cycles per second, and transmits these vibrations to the ossicles in the middle ear; these beat against the bony case of the labyrinth and stimulate the cochlea.

Hearing is effective over a wide range of pitch and intensity; it is most sensitive to frequencies around 1000 c/s. It is also direction-finding, due to the position of the ears at either side of the head and to appreciation differences in the time of arrival of sounds at either ear. Hearing impulses are transmitted along the acoustic portion of the eighth nerve.

The nose

The external part of the nose is partly fibrocartilaginous. The nasal cavity is divided into right and left halves by the septum. The two cavities open externally at the nostrils or *anterior nares* and behind into the nasopharynx at the *posterior nares*. The side wall of each cavity is marked by three ridges: the superior, middle and inferior *conchae*, the substance of each of which is a fragile scrolled *turbinate bone*; and beneath each turbinate is a recess or *meatus*. The accessory nasal air sinuses open into one or other meatus, the maxillary antrum into the middle meatus, and the frontal sinus by a more tortuous channel. The nasal mucous membrane is continuous with that of the sinuses. The smell-sensitive region is on the roof of each cavity, where the mucosa contains the olfactory cells whose fibres make up the first cranial nerve.

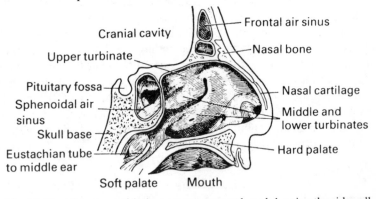

Fig. 19.19   The nose, with the septum removed, and showing the sidewall of the left nasal cavity (After *Gray*)

The sense of smell, though vestigial in man, remains one of the most delicate of the senses. It depends on odoriferous particles dissolving in the moisture of the mucous membrane. The sensory area is a patch of membrane at the summit of the nasal cavities, so there is a little delay before a new odour is appreciated. The nature of the sensation aroused probably depends on the molecular configuration of the substance. Pungent 'smells', such as that of ammonia, are really stimuli of ordinary sensation and are transmitted by a different nerve.

Taste

Taste, like smell, is a chemical sense and depends on the entry of foreign particles into solution on the tongue and palate. Microscopic *taste buds* in the mucous membrane serve this function, and in the tongue these are grouped into obvious projections, or *papillae*. Taste buds are also found on the soft palate and epiglottis. Their impulses are collected up in the seventh and ninth cranial nerves. There are four basic tastes: sweet, sour, salt and bitter, and these do not blend but remain distinguishable. The nature of any taste depends not only on the chemical composition of a substance but also on its acidity or alkalinity. Many so-called tastes and flavours are really smells, and are lost when the latter sense is put out of action during a cold.

20

The Endocrine System

Animals have two ways of responding to stimulation. The most primitive method is by an alteration of the chemical substance of the protoplasm at the stimulated point, with diffusion of the chemical agent to other parts of the body. The more sophisticated method is the development of a nervous system. However, the primitive mechanism has been retained, and even developed, in the higher animals. The *endocrine* system is a group of glands whose secretions are internal, i.e. absorbed into the bloodstream and diffused throughout the body, on whose functions they exert a profound control. These secretions are the *hormones*, or chemical messengers; and these glands differ entirely from those of external excretion, such as the liver and salivary glands, which have ducts opening into a body cavity or on the skin surface. Some organs are capable of both internal and external secretion; the pancreas secretes digestive juice into the bowel as well as insulin into the blood, and the ovaries and testicles form ova and spermatozoa as well as the male and female hormones.

A characteristic of hormones is the potency of even minute amounts. There are three main types of endocrine activity:

1 A *temporary response* to emergency, e.g. the liberation of adrenaline from the adrenal glands;
2 A *sustained activity*, e.g. the constant stimulus of the thyroid gland on metabolism;
3 The secretion of the *sex hormones* by sex organs.

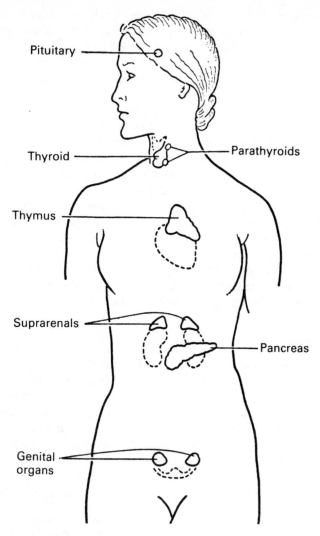

Fig. 20.1 The endocrine glands

The endocrine glands are largely, but not entirely, independent of nervous control, which is most important in relation to the adrenal medulla and the posterior pituitary. They influence each other's activities through the action of their hormones. *Chemically,* hormones are proteins (or polypeptides), such as the secretions of the anterior lobe of the pituitary; or steroids, such as those of the adrenal cortex; or phenol derivatives, such as thyroxine and adrenaline.

The activities of the endocrine glands are manifest from the clinical disturbances when their output is deficient – e.g. the cretinism of thyroid deficiency – and from the response of such individuals to treatment with gland extracts. Hormones maintain internal homeostasis by controlling water and electrolyte balance (adrenal cortex), blood-sugar (insulin) and metabolic rate (thyroid). They also influence metabolism and reproduction, growth and development, and the response to stresses. They have specific 'target' organs or tissues. These may be specific, e.g. the thyroid-stimulating hormone of the pituitary affects only the thyroid gland; or they may, like the adrenal corticosteroids, affect the metabolism of most cells.

All the endocrine glands are closely interrelated; the precise balance in the individual determines personality. They are an orchestra, of which the leader is the pituitary gland.

Pituitary

The pea-sized pituitary gland is slung beneath the brain immediately behind the optic chiasma, occupies a little pocket in the base of the skull between the two middle cranial fossae. It is attached by a short stalk to an important region of the brain called the *hypothalamus* in the floor of the third ventricle; hypothalamus and pituitary are in close circulatory communication. Although not essential to life, disease or removal of the gland results in arrest of growth, atrophy of the sexual organs, weakness and senility. It consists of an anterior and a posterior lobe, each of which produces several hormones.

Hormones of the anterior pituitary

1 *Somatotrophic or growth hormone*. This is concerned with growth, particularly of the skeleton. It is necessary for normal protein synthesis and increase in size of the tissues. Essentially it promotes anabolism. It is secreted in irregular bursts, more at night than during the day. Overproduction of this hormone, as may occur

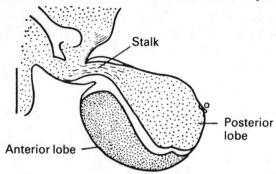

Fig. 20.2 The pituitary gland, longitudinal section (After *Gray*)

with a tumour of the secreting cells, causes *gigantism* in children due to overgrowth of the long bones. In adults, after epiphyseal closure, the corresponding condition is *acromegaly*, an overgrowth of the soft tissues of hands, feet, face and tongue, with coarsening and broadening of the features. The tumour responsible may press on the optic chiasma and cause blindness. Deficiency of growth hormone in childhood produces *pituitary dwarfs*, sexually immature individuals.

2 *Adrenocorticotrophic hormone, ACTH.* This is necessary for the activity of the cortex of the adrenal gland. No significant steroid output is possible without pituitary stimulation, and the close connection of these two glands is sometimes termed the 'pituitary-adrenal axis'. If the pituitary is destroyed the patient can be kept in health only by the administration of the adrenal hormones or of ACTH.

3 *Thyrotrophic hormone.* This is responsible for normal function of the thyroid gland.

4 *Gonadotrophic hormones.* These control the release of ova and female sex hormones by the ovary, i.e. they are responsible for the cyclical changes in the ovary (pp. 352, 354). In the male they control sperm formation and the secretion of male hormones by the testis.

5 *Prolactin.* This is essential for development of the breast during pregnancy and the secretion of milk.

Pituitary control system. The anterior lobe of the pituitary is stimulated to produce these hormones by releasing factors passed down to it from the hypothalamus, and the hypothalamus is under the control of higher cerebral centres in response to various stresses. After the appropriate pituitary hormone has reached the target organ – say, the adrenal cortex – the latter is stimulated to secrete its own hormone – in this case, cortisol – into the bloodstream. An ingenious feedback mechanism now comes into play. Once the blood level of cortisol has reached an optimum, this inhibits further production of ACTH by the pituitary. Then, as the concentration of target gland hormone falls, the pituitary is reactivated and produces more of its trophic hormone. This mechanism applies to all the secretions of the anterior pituitary and is represented below:

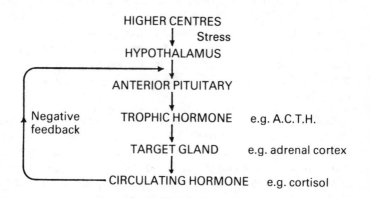

HIGHER CENTRES
↓ Stress
HYPOTHALAMUS
↓
ANTERIOR PITUITARY
↓
Negative feedback   TROPHIC HORMONE   e.g. A.C.T.H.
↓
TARGET GLAND   e.g. adrenal cortex
↓
CIRCULATING HORMONE   e.g. cortisol

Hormones of the posterior pituitary

1 *Antidiuretic hormone, ADH.* This controls water excretion by the kidney (see p. 300). It increases the permeability to water of the terminal ducts of the renal tubules, and so promotes the reabsorption of water into the circulation and concentrates the urine. Again, a feedback mechanism comes into play, designed to maintain constant osmolarity in the blood and extracellular fluid. Should the osmotic pressure of the plasma rise, this stimulates specific osmoreceptors in the brain, more ADH is liberated by the posterior pituitary, water is retained in the blood to exert a diluting effect and a smaller volume of concentrated urine is passed. Should the osmotic pressure of the plasma fall, ADH secretion is inhibited, a

larger amount of dilute urine is passed and the plasma becomes more concentrated.

2 *Oxytocin*. This exerts a stimulating action on smooth muscle, notably of the uterus and the milk ducts in the breast. Oxytocin plays an important part in uterine contraction during childbirth and in ejection of milk from the breast.

Thyroid

The thyroid has two main secretions:

1 *Thyroxine*. This is secreted in response to the stimulus of the thyrotropic hormone of the pituitary. It is a complex glycoprotein containing iodine. Dietary iodine is derived from the soil, and where iodine is deficient the thyroid hypertrophies to try to compensate and forms a swelling, or *goitre*, in the neck. The condition is prevented by iodising table salt.

Thyroxine stimulates metabolism; it increases the rate of tissue build-up and breakdown, and consumption of oxygen, and is essential to normal development. Excess of secretion causes *thyrotoxicosis* or Graves's disease – overexcitability, rapid pulse, staring eyes and wasting. Deficient secretion in children causes *cretinism* – dwarfism and mental retardation – and in adults *myxoedema*, an obese slow-wittedness.

2 *Calcitonin*. This transfers calcium from the plasma to bone. It is used in the treatment of conditions characterised by rarefaction of bone.

The *parathyroids* are four small pea-like glands embedded in the thyroid capsule. Their hormone acts together with calcitonin to regulate the level of plasma calcium at 9–11 mg/100 ml. Parathormone mobilises calcium from bone into the blood and promotes intestinal absorption of calcium. It is secreted whenever plasma calcium tends to fall; should the latter rise, calcitonin comes into play. If there is an excess of parathormone, as with a parathyroid gland tumour, so much calcium is moved out of the bones that they appear transparent in the X-ray, and fracture easily. If there is not enough parathormone, as may happen when the glands are inadvertently removed, the plasma calcium falls to low levels, causing convulsions.

Adrenal (suprarenal) glands

These are triangular, cap-like organs sitting on the upper poles of the kidneys. They have an outer rind, or *cortex*, and an inner *medulla*.

The function of the *medulla* is intimately connected with that of the sympathetic system. Its hormones – *adrenaline* and *noradrenaline* – are secreted in response to sympathetic stimulation in emergency. Their release excites the usual adrenergic effects at the termini of the sympathetic system. There is an increase in the rate and force of the heartbeat, the blood-pressure rises, the smooth muscle of bowel and bladder relaxes, the pupils dilate and the hair stands on end; blood is diverted to skeletal muscle and glycogen mobilised from the liver to prepare for fight or flight.

The internal secretions of the adrenal cortex are known as corticosteroids and are based on the carbon ring structure:

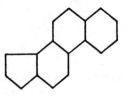

The most important are cortisol (hydrocortisone) and aldosterone, which have a profound influence on the metabolism of carbohydrates and electrolytes respectively. *Cortisol* promotes glucose formation and retention, favours protein breakdown and inhibits tissue repair; *aldosterone* modifies the transport of sodium and potassium ions across cell membranes, favouring sodium retention and potassium excretion in the renal tubules.

Cortisol formation is increased in stress situations – not just emergencies but also those of a more prolonged nature after operations and injuries, in convalescence and emotional crises. It is produced in response to stimulation of the adrenal cortex via the hypothalamic-pituitary-adrenal axis, and if the cortex is unable to respond the individual may collapse and die.

The adrenal cortex also has a close connection with the sex glands. Abnormal function of the adrenal cortex can bring on precocious sexual development or actually tend to *reverse* the sexual characteristics, producing masculinity in women and feminism in men. Some unfortunate women looking like men and with male sexual desires, are the victims of an overacting tumour of the adrenal cortex, and some have been cured by its surgical removal.

Pancreas and insulin

Insulin is the internal secretion of groups of cells in the pancreas known as the *islets of Langerhans*. It is an anabolic hormone that promotes the storage of carbohydrate, protein and fat. Insulin is secreted in response to the rise in blood sugar after a meal and acts to lower this level by facilitating the take-up of glucose in the tissues and inhibiting the formation of glucose in the liver. It also promotes glycogen deposition in muscle.

The *thymus* may be regarded as an endocrine gland, inasmuch as it secretes a factor essential for the development of lymphoid tissue in the lymph nodes and spleen in the newborn, and the maintenance of these tissues and the production of new immunologically competent lymphocytes in the adult.

The gonads or sex organs

The ovaries and testes, besides forming the reproductive cells – ova and spermatozoa – secrete the female and male sex hormones respectively, *oestrogens* and *androgens*. These are steroid hormones, formed in response to gonadotrophic drive from the anterior pituitary, with the usual negative feedback control.

These hormones are responsible for the development as the *secondary sexual characteristics* – the attributes, though not the essentials, of sexuality. These include hairiness, a deep voice, enlargement of the external genital organs and a narrow pelvis in men; hairlessness, smooth skin with plenty of subcutaneous fat, the development of the breasts and the broad pelvis of women.

In more detail (see also Chapter 21 on reproduction), the ovarian hormones consist of two groups: oestrogens and progestagens. Oestrogens initiate the cyclic activity of genital tract and breast, promoting thickening and secretion in the lining membranes, and muscular development in the walls of uterus and vagina. They are more concerned with the first, preparatory or follicular, stage of the menstrual cycle associated with ovulation. The progestagens have to do with the second, receptive or luteal, half of the cycle, in readiness for possible pregnancy, and their activity persists if pregnancy ensues.

Oestrogen also has some general metabolic effects. Its production falls off sharply at the menopause, often resulting in some emotional instability and masculinisation, both of which can be relieved by giving the hormone.

The most important androgenic hormone is *testosterone*. This stimulates the growth and function of the male reproductive tract. It also has a marked anabolic action on general metabolism, promoting protein synthesis, growth and skeletal development.

Finally, and paradoxically, ovary and testis also secrete small amounts of the *opposite* sex hormones.

21

Reproduction and Development

Reproductive organs

The essential sex organs are the *gonads*: a pair of *testes* in the male forming the spermatozoa, a pair of *ovaries* in the female forming the ova. The general plan of the reproductive system in both sexes is not dissimilar, but the ovaries remain in the abdominal cavity while the testes lie outside it. And in the female there is a uterus to house the developing embryo, while the penis of the male is represented in the female by the diminutive clitoris.

Male organs

The *testes* develop in the abdominal cavity but pass down before birth to enter the loose skin pocket of the scrotum externally, where they hang down on each side of the root of the penis. There remains an oblique passage through the abdominal muscles just above the inguinal ligament, the inguinal canal, occupied by the stalk of the testis or *spermatic cord*, which carries blood vessels to the organ and the sperm duct back to the pelvis.

The testes are ovoid, with a tough capsule made up of lobules containing the fine tubes in which the spermatozoa are formed. Applied to their outer side is a curved organ, the *epididymis*, which receives sperm from the testis; it is an intricately coiled tube from which issues the main spermatic channel, the *vas deferens*, running up to the abdomen. Testis and epididymis lie vertically in the scrotum, surrounded by a loose serous sac.

The vas deferens runs down the side wall of the pelvis to reach the

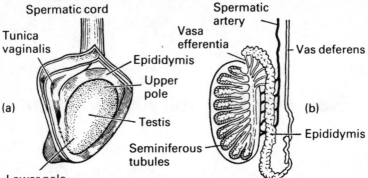

Fig. 21.1 (a) Testis and epididymis exposed by reflection of their serous covering, the tunica vaginalis. (b) The same in longitudinal section

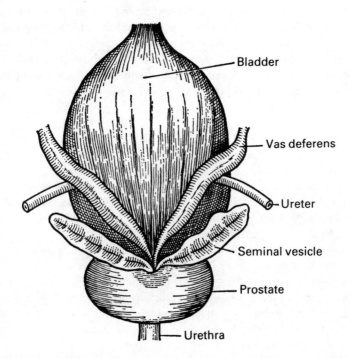

Fig. 21.2 Bladder and associated organs in the male (posterior aspect)

back of the lower part of the bladder, where it lies on the upper surface of the prostate; here a sac is attached to its outer side, the *seminal vesicle*, for the storage of sperm. The ducts of vas and vesicle open via a common·*ejaculatory duct* into the *urethra*, which lies embedded in the prostate gland. Sperm are continuously formed in the testis, stored in the vesicle and only enter the urethra in the ejaculation of orgasm, when they are discharged via the urethra and penis into the vagina. The ejaculated fluid is not just the sperm but a complex fluid containing also the secretions of vesicles and prostate. The *prostate* is a solid organ of muscular and glandular tissue, the shape and size of a chestnut. It has its base applied to the neck of the bladder and its apex on the pelvic floor, and is traversed by the urethra.

The *penis* consists of a central *bulb* arising from the centre of the perineum, and traversed by the urethra and two lateral *crura* springing from the sides of the pubic arch. These join to form the shaft of the organ, a cross-section of which is shown in Fig. 21.3. Here, the part containing the urethra, the continuation of the bulb, is the *corpus spongiosum* below, with the *corpora cavernosa*, continuations of the two crura, above on each side. It is to these latter that the organ owes its property of increase in length and girth on sexual excitement, becoming rigid and capable of introduction into the vagina. This process of erection is due to a system of cavernous

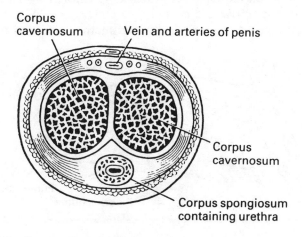

Fig. 21.3 Cross-section of shaft of penis (After *Gray*)

spaces which can be rapidly distended with blood from the penile arteries.

Female organs

The *ovaries* are a pair of almond-shaped organs, lying on the sidewall of the pelvis just below its brim. They are studded with fluid-filled cysts, the ovarian *follicles* in which the egg cells or *ova* ripen, one coming to maturity every month. The ovaries are embedded in the *broad ligament*, which stretches from the uterus to either side of the pelvis. In the upper free edge of this ligament are the *uterine tubes*, attached to the uterus like outstretched arms. These muscular channels open into the uterus medially and have at their outer ends a fringed, funnel-like entrance, which embraces the ovary so as to receive the ovum when it is shed.

Uterus. The womb is a hollow organ, with thick muscular walls, lying in the pelvis between bladder in front and rectum behind (Fig. 9.1). It communicates below with the vagina and at each side with the uterine tubes. It is tilted forwards so that the anterior surface rests on the bladder; both surfaces and the dome (or *fundus*) are covered with peritoneum, and the peritoneal pouch between uterus and rectum is the deepest part of the abdominal cavity.

The virgin uterus is 7·5 cm long, its wall 2·5 cm thick. Longitudinal section shows the upper *body*, 5 cm long, with a triangular cavity, and the lower *cervix*, 2·5 cm in length and traversed by a narrow canal opening into the main cavity at its internal orifice and into the

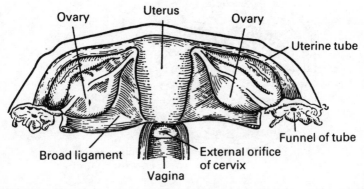

Fig. 21.4   Internal sex organs of the female (After *Gray*)

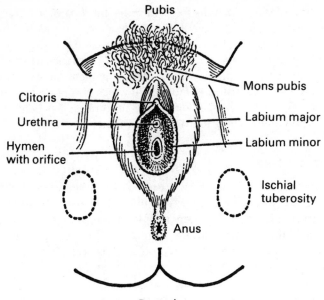

Fig. 21.5 The female perineum

vagina at its external orifice. The lining mucous membrane of the body, or *endometrium*, contains mucous glands and undergoes striking changes during the menstrual cycle. The organ becomes enormously enlarged in pregnancy but returns almost to normal after childbirth. The uterus is small and undeveloped in childhood and atrophies in old age.

The *vagina* is a distensible canal, capable of receiving the penis during intercourse and of allowing passage of the child in parturition. It extends from the uterus, which it meets at an angle of 90°, to run down and forwards through the pelvic floor and to open externally on the perineum. Part of the cervix protrudes into the vaginal vault, the encircling rim of which is known as the fornix. The front wall of the vagina is blended with the back of the bladder and the urethra; the back wall is separated from the rectum by fibrous tissue.

The *external female genitalia* are known as the *vulva*. This includes the *mons pubis*, a fatty hair-covered eminence in front of the pubic symphysis; the skin folds forming the lips of the vaginal orifice, comprising the outer thick *labia majora* and the inner slender *labia minora*; the *clitoris*, a diminutive but sensitive erectile equivalent of the penis, lying at the meeting of the labia in front; and the *hymen*, an incomplete partition of mucous membrane stretching across the vaginal orifice in the virgin, which is ruptured by intercourse. The external opening of the urethra is just behind the clitoris and immediately in front of the vagina.

Reproduction

In animals, sexual desire usually alternates with seasons of indifference. Even when the male is continuously active, the female is receptive only in cycles, dependent on intermittent rhythm of activity in the ovaries, and the formation of the sex hormones is discontinuous. In humans, sexual pressure is pretty constant, particularly in the male, as is secretion of sex hormones. There is still some periodicity in the female, associated with the monthly phenomenon of *menstruation*, a flow of blood and mucus from the womb which occurs only in the higher apes and man.

Development of the sexual organs before puberty and their functioning thereafter depend on the pituitary gland. But when puberty arrives it is the hormones of the sex organs themselves that cause the development of the secondary sexual characteristics and the origin of sexual desire, in the girl the swelling of the breasts and the commencement of menstruation, and in the boy the change of voice and enlargement of the external organs. All these changes depend on the primary development of the ovaries and testes.

The menstrual cycle

This forms a complicated pattern, regulated by both pituitary and ovarian hormones. It is usually considered in terms of a twenty-eight-day cycle beginning with the first day of menstrual bleeding, though longer and shorter cycles are common.

In the twenty-eight-day cycle *ovulation* occurs at or around the fourteenth day, i.e. an ovarian follicle ripens and discharges the ovum into the peritoneal cavity. This egg cell is taken up by the uterine tube and propelled by muscular peristalsis towards the uterus. Here, the lining has become engorged and thickened in preparation for the embedding of a fertilised ovum, for fertilisation, if it has occurred, will have taken place in the tube. If, as is usually the case, the ovum has not been fertilised, the mucous membrane of the uterus is shed in the bleeding of menstruation. This lasts three to five days, and then a new lining is built up and is completed by the fourteenth day of the cycle, midway between two periods, just when ovulation occurs again. Thus the cycle affects ovaries and uterus and is divided into the *follicular phase* – the first two weeks, and the *luteal phase* – the second fortnight.

The hormonal aspect of all this is as follows. The ripening of the ovum in its follicle and its discharge at ovulation are under the control of the anterior pituitary hormone – the gonadotrophin known as FSH (follicle-stimulating hormone). Once the egg has left the ovary, the remains of the follicle become organised into a bright yellow structure known as the *corpus luteum* under the influence of another pituitary gonadotrophin – LH or luteinising hormone. The follicle itself has a hormone which stimulates thickening of the uterine lining preparatory to possible conception, and the corpus luteum has a hormone which completes these preparations. Should pregnancy supervene, the corpus luteum enlarges and persists; otherwise it disintegrates. Thus ovulation and menstruation alternate at fortnightly intervals; the uterine lining is completed and receptive during the latter half of the cycle, destroyed at menstruation, and rebuilt in the first half.

Fertilisation

Union of the male spermatozoön with the female ovum normally occurs in the uterine tube as the egg is moving towards the uterus. It can only occur if a living sperm derived from the male by recent intercourse has made its way up the vagina, through the cervix and body of the uterus into the tube. Success therefore depends on a near coincidence of intercourse and ovulation, say within forty-eight hours, so that the likelihood of any single sexual act resulting in pregnancy is small. When the male ejaculates, a pool of semen is deposited in the vaginal vault around the cervix. The sperm have to penetrate the cervical mucus and progress upward at not more than 3 mm/min; their overall progress in the uterus and tube is much faster, due to muscular propulsion by these organs. During their climb there is an enormous reduction of sperm count. Around 100

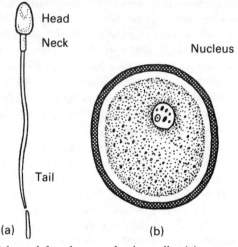

Fig. 21.6 Male and female reproductive cells, (a) spermatozoön, (b) ovum

million spermatozoa are deposited in the vagina, but only a million enter the uterus and perhaps only 100 reach the ovum. Spermatozoa do not remain active and fertile in the female tract for more than two days. Once the ovum has been penetrated by a spermatozoön, rapid changes ensue. The nuclei of the two cells fuse and the ovum becomes impenetrable to further spermatozoa.

The sperm cell has a flattened pear-shaped head containing the nuclear material, connected by a short neck to the long tail, whose lashing movement gives the cell its mobility. Sperm formation in the testis is a complicated process and many spermatozoa are ill-formed and unsuitable for fertilisation. If sperm are very few or even totally absent from the seminal fluid this is one possible cause of infertility. Spermatogenesis is also impaired if the scrotum is kept too warm.

Sterility may be due to the inability of a woman to house a developing embryo within her uterus, but it can be due to an error in either sex. The sex organs may be abnormal, diseased or absent; the sperm may be few, malformed or non-viable; the vaginal secretion may be lethal to them; the uterine tubes may be blocked. In some cases, sterility is due to failure of proper maturation of ova, and this can be treated by administering the pituitary follicle-stimulating hormone, though with some risk of multiple births.

A point of great importance in the development of male and female germ cells is that, at some stage, their chromosome count has to be reduced. A normal body cell contains forty-six chromosomes, which is sometimes called the *diploid* state, and if this applied to ovum and spermatozoön the fertilised egg would contain ninety-two chromosomes. Hence, there is a cell division at a stage in germ cell formation when the chromosomes, instead of splitting in two as they usually do, are simply divided between the two new cells, the final germ cell containing twenty-three chromosomes – the *haploid* state.

Life cycle in male and female
Adolescence describes a state of physical and social maturity, and cannot be allotted any definite time. *Puberty* means the capacity for sexual procreation – in the female that ovulation has begun and in the male that emissions of spermatozoa are occurring. The age at which regular uterine bleeding begins is called the *menarché*, and this occurs at an average age of thirteen, but ranging between ten and seventeen years. The onset of puberty is a complex hormonal interaction; hypothalamus, anterior pituitary, ovaries and adrenal cortex all play their part.

The cessation of menstruation is known as the *menopause*, which occurs between the ages of forty and fifty-five. Many bodily and mental changes, known as the climacteric, are associated with this. The ovaries cease to produce either ova or oestrogen, the genital tract and breasts atrophy, with widespread physiological and psychological disturbance.

In the male, sperm formation begins after the age of ten or eleven, and with this occurs enlargement of the penis and scrotum and the later appearance of pubic hair. The voice breaks and the sweat and sebaceous glands overact, which is why acne is so common. Skeletal growth accelerates, and this growth spurt lasts until the closure of the epiphyses of the long bones, a year or two later than in girls – hence the greater average height of men. It is difficult to pinpoint the onset of puberty in boys because there is no obvious guideline such as menstruation. It is usually between ten and fifteen years of age. The basic mechanism of hypothalamic-pituitary-adrenal-gonadal control is similar to that of the female. Once established, testicular function continues for the rest of life and is only slightly diminished in old age. There is no sharp cut-off of gonadal function as in women. Nevertheless, at about the corresponding age many men do suffer some waning of desire and physiological disturbance, suggesting that there is such a thing as a male climacteric.

Development of the embryo

The nuclei of the two germ cells fuse, and the egg begins to divide to form a rounded mass of cells, whose number doubles at each division. This process begins before entry into the uterus, the journey along the uterine tube taking several days. Once in the uterine cavity, the ovum adheres like a parasite, excavating a cavity in the mucous membrane, in which it becomes embedded.

The early embryo consists of little more than a pair of sacs or vesicles, the *amniotic cavity* and the *yolk sac*, with an intervening plate which is the actual embryonic area at which development

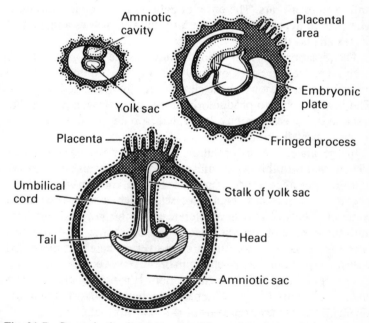

Fig. 21.7 Stages in the development of the early embryo

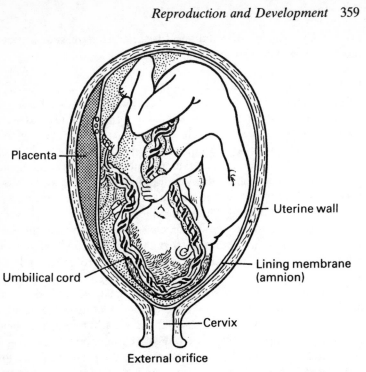

Fig. 21.8 The foetus *in utero* (After *Gray*)

occurs. The amniotic cavity soon greatly expands, coming to line the whole of the uterus, and is filled with the amniotic fluid bathing the embryo. At the site of implantation, the amniotic membrane develops a complex system of fringed processes penetrating the uterine wall. This constitutes the mass of the after-birth or *placenta*, a fleshy disc of 15–20 cm diameter when maximally developed. It is here that maternal and embryonic circulations are in closest proximity; but there is no direct communication; oxygen and nutrients must diffuse across the separating layer of cells. The yolk sac becomes squeezed by the growth of the amnion into a narrow stalk, the core of the *umbilical cord* which connects the developing child to the placenta. The cord is 51 cm long at birth, spirally twisted, and carries in its jelly-like substance the two umbilical arteries and the single umbilical vein of the foetus.

Since the foetus sends its stale blood to the placenta for oxygenation and nutriment, the umbilical arteries contain stale venous blood and the veins fresh arterial blood.

Embryonic layers

The early embryo soon comes to possess three layers of cells.

The outer layer or *ectoderm* forms the skin, its glands, hair and nails, the nervous system, and the essential parts of the eye, ear and nose. The nervous system originally lies on the surface of the back of the body and is infolded in development as the neural tube, the precursor of the spinal cord, with a bulbous expansion at the head end marking the brain. All the sense organs originate in this layer; an infolding of modified skin surface meets an outgrowth from the central nervous system. This arrangement is well shown in the eye, where the retina, an extension from the brain on the stalk of the optic nerve, cups the transparent lens developed from the overlying skin.

The innermost layer or *endoderm* forms the bowel and its associated glands – such as liver and pancreas – the lining of the respiratory tract and lungs, and the thyroid and parathyroid glands. The intestine develops on the back wall of the abdominal cavity and comes to lie suspended by a mesentery still attached to the posterior wall. The large bowel undergoes a complicated rotation to bring the caecum and ascending colon to the right and the descending colon to the left.

The intermediate layer is the *mesoderm*, origin of the connective tissues of the body: bone and cartilage; the teeth; the muscles, both voluntary and involuntary; the heart and blood vessels; and the urogenital system.

At any point in the complex pattern of development things may go wrong, or not go far enough. The spinal cord may be left on the surface of the back as an open layer; or it may develop normally, but the two halves of the vertebral column fail to enclose it completely, leaving a gap behind, the condition of *spina bifida*. The palate may remain cleft if its two halves fail to meet; the septum of the heart may remain incomplete, allowing a disabling admixture of venous with arterial blood; a bone or a limb may fail to develop wholly or in part; fingers and thumbs may be too few or too many. What is outstanding is that any error is a rarity.

Time scale

The two primitive vesicles of the embryo are present ten days after fertilisation and the intervening embryonic area begins to develop in the third week; head and tail folds, neural groove and heart are obvious by the fourth week. In the fifth week appear the lens of the eye, the rudiments of the face, the gill arches and the stumps of the limb-buds, and the embryo is now 5 mm long. The sixth week sees the body curved on itself, the head approximating to the long tail, the umbilical cord attached to the belly near the latter; the liver enlarges, and the limb-buds grow out and are demarcated into segments.

By the end of the eighth week, the embryo is 2·5 cm long and is now known as the *foetus*; eyes, ears and nostrils are formed, the external genitals differentiated, and the fingers and toes marked out. Fine downy hair appears in the fourth month when the foetus is 20 cm long; in the fifth month, foetal movements begin and the skin becomes covered with a greasy secretion.

By the seventh month, the eyelids have opened and the testicles descend into the scrotum; and though the foetus may now be viable (i.e. capable of living), if born prematurely it is not adequately clothed with subcutaneous fat until the end of the ninth month.

At *birth*, the uterus contracts, rupturing the membranes and expelling first the foetus and amniotic fluid, then the placenta and finally the membranes. Certain important changes occur in the child at birth. During foetal life, the lungs have remained unexpanded and airless; they need no blood, and the stream in the pulmonary artery is shunted through a bypass directly into the aorta. With the first cry, the lungs expand, the pulmonary circulation is established and the bypass closes off. The portions of the umbilical vessels contained within the foetus also become obliterated, and the parasite has become an independent organism.

The breasts and milk

The mammary glands, designed for milk secretion, form two large, rounded eminences between the skin and deep fascia on the front of the chest, overlying the pectoral muscles; they extend from the second to the sixth ribs and from beside the sternum to the axilla. Small before puberty, they develop with the uterus, hypertrophy in

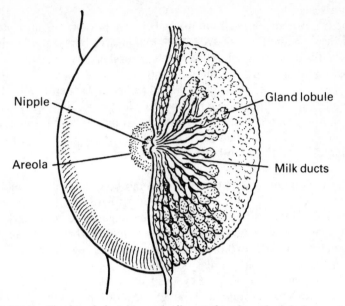

Fig. 21.9   The female breast, partly dissected to show the lobules

pregnancy – and especially at lactation after delivery – and atrophy in old age. There are minor cyclical changes with menstruation.

The pigmented nipple at the apex of the breast is surrounded by a coloured area of skin, the *areola*; this is studded with sebaceous glands, is pink in the virgin and permanently darkened in the first pregnancy. The nipple is perforated by fifteen to twenty milk ducts opening at its summit; it is sensitive and contains erectile tissue. The gland is divided into lobes by fibrous partitions, some of which fasten the organ to the chest wall. Each lobe contains a branched system of secreting gland spaces supported by fatty tissue; the larger ducts converge on the nipple and expand just before reaching it as the milk sinuses or reservoirs. The gland spaces are only distended with milk during active functioning.

The breasts develop throughout pregnancy in preparation for lactation. The first few days after delivery they yield only a watery fluid or colostrum containing a few fat globules; only after this does milk secretion begin properly. This is governed by the hormones of pituitary and ovaries, and there is also a reciprocal relation between lactation and menstruation; resumption of the periods after delivery may be delayed by continued suckling. In addition, suckling is a powerful stimulus to the involution of the uterus after childbirth.

The milk is a slightly acid fluid, opaque from the presence of finely dispersed fat globules. Its main constituents are:

1   *Protein* (caseinogen), which is converted into a solid casein clot or curd in the stomach; from this can be expressed a liquid whey;
2   *Sugar* (lactose);
3   *Fat*.

In man these are found in the proportions of protein 1·5%, lactose 6·5% and fat 3·5%; cows' milk contains more protein and less sugar, the fat content being much the same. It can be approximated to human milk by adding water to dilute the protein content and making up the sugar; but it is a second-best substitute, and the benefits natural feeding confers on both mother and child should make breast feeding the first choice.

Index

HUMAN BIOLOGY

DERYCK TAVERNER

A fully illustrated and highly readable introductory text which
provides a fascinating guide to the human body and how it works.
The human body consists of more than one trillion cells, organised
and coordinated to form a functional whole. This book explains
how these cells come together to form tissues, organs and systems,
and how the balance of the whole sensitive and complex organism
is maintained.

Dr Taverner deals first with basic body chemistry and the types of
cells and tissues, and then looks in turn at the skeletal system, the
alimentary system, the respiratory system and the nervous system.
He provides a straightforward account of the composition of blood
and the circulatory system, and explains the working of the kidneys,
lungs, heart and brain before going on to describe the senses,
glands and reproduction, and the physiology of the human life
cycle. The emphasis throughout is on the living organism in an
environment to which it is continually responding and adapting.

TEACH YOURSELF BOOKS